Operationalizing VMware NSX®

Kevin Lees, Chief Technologist for
IT Operations Transformation, VMware

Foreword by Bruce Davie, Vice President & CTO, VMware Asia, Pacific, Japan

Thanks James!

Kevin Lees

vmware PRESS

VMWARE PRESS

Program Managers

Katie Holms
Shinie Shaw

Technical Writer

Rob Greanias

Graphics Manager

Elaine Tai

Production Manager

Sappington

Warning & Disclaimer

Every effort has been made to make this book as complete and as accurate as possible, but no warranty or fitness is implied. The information provided is on an "as is" basis. The authors, VMware Press, VMware, and the publisher shall have neither liability nor responsibility to any person or entity with respect to any loss or damages arising from the information contained in this book.

The opinions expressed in this book belong to the author and are not necessarily those of VMware.

VMware, Inc. 3401 Hillview Avenue Palo Alto CA 94304 USA
Tel 877-486-9273 Fax 650-427-5001 www.vmware.com.

Table of Contents

Preface XIII
Foreword XIV
Chapter 1 - Introduction 1
What does "Operationalizing NSX" mean 2
Why IT should adjust the way it operates for SDDC/NSX 2
Who the guide is for 3
What it will teach 4
Why it matters 4
How to proceed 4
Chapter 2 - Measuring Progress 7
Why measure 8
How and what to measure 8
Finding quick wins and with whom to share results 8
Chapter 3 - People Considerations 11
Why SDDC (and NSX) changes the people equation 12
Team construction 13
Roles & skillsets 17
Culture & mindset 25
KPIs to measure progress 26
Chapter 4 - Process Considerations 29
Intelligent Operations 29
Proactive monitoring for performance and availability 30
Proactive capacity monitoring and planning 32
Change management 33
Configuration management 36
Provisioning NSX capabilities 37
Incident Management 41
Compliance management 42
Chapter 5 - Consuming NSX 49
On-boarding consumers, applications, and services 50
Applications 52
Services 54
Developer consumption models 55
Chapter 6 - Tools for Monitoring & Troubleshooting 57
NSX Native Tools 58
vSphere Tools 60
VMware vRealize Intelligent

Operations Tools ...61
vRealize Network Insight ...61
vRealize Operations ... 66
vRealize Log Insight ... 66
Partner Ecosystem Tools ... 69
Chapter 7 - Conclusion ... 71
Where to go for more information ...72
Index ...75

List of Tables

Table 3.1 Architecture roles ...18
Table 3.2 Engineering roles ... 20
Table 3.3 Administrator roles ...23
Table 3.4 Sample team structure KPIs ...26
Table 4.1 Sample proactive performance and availability monitoring KPIs ...32
Table 4.2 Sample proactive capacity monitoring and planning KPIs ...33
Table 4.3 Sample change management optimization KPIs ...36
Table 4.4 Sample configuration management optimization KPI ...37
Table 4.5 Sample integrated NSX provisioning optimization KPIs ...40
Table 4.6 Sample incident management impact KPIs ...42
Table 4.7 Sample compliance management impact KPI ... 46
Table 5.1 Sample on-boarding impact KPI ...52
Table 5.2 Sample application on-boarding impact KPI ... 54
Table 6.1 Sample Native Tools troubleshooting capabilities ...59
Table 6.2 Sample vSphere-based troubleshooting capabilities ...60
Table 7.1 Reference ...72

List of Figures

Figure 1.1 Typical beginning state and target end state.......... 3
Figure 3.1 Cross-domain and cross-functional silos.......... 13
Figure 3.2 Blended team.......... 14
Figure 3.3 Getting started.......... 15
Figure 4.1 Example of vRealize Network Insight's 360° visibility.......... 31
Figure 4.2 Example of micro-secgmentation in a multi-application deployment.......... 35
Figure 4.3 vRealize Automation and NSX.......... 38
Figure 4.4 vRealize Automation and NSX integrated capability.......... 39
Figure 4.5 Distributed Firewall Event Summary in vRealize Log Insight... 44
Figure 4.6 Auditing for access attemps using vRealize Log Insight.......... 44
Figure 4.7 Firewall rule membership changes over time in vRealize Network Insight.......... 45
Figure 4.8 Creating change alerts for auditing in vRealize Network Insight.......... 46
Figure 5.1 Separation between dev/test and production environments.. 50
Figure 5.2 Isolated Tenant environments.......... 51
Figure 6.1 NSX tool ecosystem.......... 58
Figure 6.2 360° topology view.......... 62
Figure 6.3 Example Palo Alto Networks VRF configuration.......... 62
Figure 6.4 Example best practice checklist failure listing.......... 64
Figure 6.6 Customer best practice checklist-based problems detected.. 65
Figure 6.7 NSX Overview dashboard in vRealize Log Insight.......... 67
Figure 6.8 Distributed Firewall Overview dashboard in vRealize Log Insight.......... 67
Figure 6.9 Firewall Actions log details in vRealize Log Insight.......... 68

About the Author

Kevin Lees is the field Chief Technologist for IT Operations Transformation at VMware, focused on how customers optimize the way they operate VMware-supported environments and solutions. He is responsible for defining, communicating, and evangelizing VMware's IT Operations Transformation vision and strategy as it relates to operational (integrated organization, people, process, and application of VMware technology) approaches and best practices. Kevin also serves as an advisor to global customer senior executives for their IT operations transformation initiatives. Additionally, he leads the IT Transformation activities in VMware's Global Field Office of the CTO.

Kevin is an industry-recognized IT Operations Transformation thought leader, strategist, and evangelist. He was the global services delivery team lead for VMware's Cloud Practice from 2009 through mid-2011 and led many of VMware's early cloud implementations with Fortune 100 customers. Since 2011, Kevin has focused on initiating, building, and supporting IT Operations Transformation as a consultative delivery practice within VMware. Kevin has a combined 35+ years of IT system architecture, design, and integration, IT Operations, and IT consulting experience.

Kevin Lees can be reached at klees@vmware.com

Content Contributors

Vyenkatesh (Venky) Deshpande works as a Sr. Product line manager in the Networking and Security Business Unit at VMware. Venky focuses on the operational aspects of the NSX platform and drives the product requirements and partnership effort with the eco system. He has helped many NSX customers evolve their organizations from People and Process point of view such that they are successful in operationalizing NSX. Venky has more than 15 years of experience in the networking industry and has expertise in building products and solutions for the campus, wan and data center networks.

Mark Schweighardt is Director of Product Marketing in the Networking & Security Business Unit at VMware. For the past 17 years, Mark has held various product management and product marketing positions, mostly at Silicon Valley start-ups focused on solving critical IT security problems. Mark has worked closely with hundreds of large enterprise companies from a variety of industries, including financial services, banking, healthcare, high tech, and many other industries. Prior to joining VMware, Mark held positions at Voltage Security, Encentuate, and ActivIdentity.

Hammad Alam is a Lead Solutions Architect within VMware's Networking and Security Business Unit working for Customer Success Organization. Hammad's current focus is to look at NSX from technology standpoint and extend it to successful adoption in organizations. Architecting, advising and evangelizing NSX from People, Process and Technology perspectives to provide the holistic picture needed for any new-ground breaking technology to be successful in an organization.

Additional Contributors:

Bode Fatona

Neil Mansukhani

Paul Wiggett

Chris McCain

Bill Erdman

Acknowledgements

It takes the knowledge and resources of multiple individuals to successfully create a guide like *Operationalizing VMware NSX*. I would like to thank the following people for their support in developing and reviewing the material included:

A special thanks to the VMware Networking and Security Business Unit's product marketing team, including Shinie Shaw and Katie Holms (Program Managers) for your direction, encouragement, and drive in creating this book.

Thank you Rob Greanias (Technical Writer) and Elaine Tai (Graphics Manager) for your efforts in completing this book.

Thank you Hammad Alam, Mark Schweighardt, and Vyenkatesh Deshpande – without whose contributions and insights this book would never have come to fruition.

Thank you Bode Fatona, Neil Mansukhani, Paul Wiggett, and other members of the PSO field team for your input and review, and for letting me leverage your wealth of experience implementing IT operational transformations for organizations worldwide.

Thank you Chris McCain of the VMware Networking and Security Business Unit, along with Bill Erdman of the VMware Cloud Management Business Unit, for your critical input and review.

Thank you, Kausum Kumar, Romain Decker, Chris Kunselman, Jose Alamo, Geoff Wilmington, and the many others whose work provided the inspiration for different aspects of the book.

Preface

Operationalizing VMware NSX® offers guidance to management-level decision makers and influencers concerned with the impact of VMware NSX on their organization, staff, and operational processes. It also speaks to engineers and technical decision makers responsible for optimizing operational tooling in a VMware NSX environment.

Operationalizing VMware NSX provides the information needed to optimize the on-going operations of a VMware NSX environment. Specific areas examined include the critical aspects of team structure, culture, roles, responsibilities, and skillsets. Additional guidance is provided for operational processes, monitoring, and troubleshooting in a VMware NSX environment.

Foreword

The idea of network virtualization has been around since at least the early 2000s, but commercial adoption of the technology really took off around 2012. The launch of Nicira's "Network Virtualization Platform (NVP)" and subsequent acquisition of Nicira by VMware brought network virtualization to a broad audience. In 2013, VMware formally launched NSX™, the network virtualization platform, which is now an increasingly common choice for the delivery of networking and security services.

As it happened, I joined Nicira just before the formal launch of NVP, and I have witnessed the early adoption of NSX through to increasingly mainstream acceptance today. One clear trend over the last five years has been a shift from talking about the architecture of network virtualization to a focus on the operational aspects. In the early days, customers were looking to be convinced that our approach was technically valid, but as adoption accelerated, the discussion shifted to operational issues such as trouble-shooting, upgrades, maintenance, automation, integration with other products and third-party tools, and so on. As a product team, we increasingly focused on building operational capabilities into the NSX product and on exposing APIs that would allow other tools to interact with NSX to provide operational support.

NSX is not like other networking products that preceded it. Networking delivered in software opens up the possibility of moving more operational tasks into software. It also requires organizations to rethink how they operate their network, as it is no longer an independent silo of hardware devices but an integrated component of the software-defined data center. These organizational changes can be harder to achieve then a simple change in technology.

Which brings me to this book. Operationalizing NSX is essential if a customer is to benefit from the capabilities of network virtualization. As with many disruptive technologies, the true benefits accrue when processes change to make most use of new capabilities. Kevin Lees has spent his career helping customers transform their IT operations, and now brings his focus to this timely topic. This book brings together the lessons learned over five years of changing the way networks are built and operated, and should be read by anyone who is serious about deploying NSX and realizing the many benefits it can bring.

Bruce Davie, CTO, Asia Pacific & Japan

Chapter 1

Introduction

The Software Defined Datacenter (SDDC) presents a nearly unprecedented opportunity for IT to dramatically increase agility and improve time to value in support of business initiatives. By introducing software defined networking and security, VMware NSX™ has become a core and very impactful component of the SDDC – but like the SDDC is an enabling technology. As an enabling technology, simply implementing NSX from a technology perspective alone does not guarantee IT will become more agile or increase the speed in delivering business value.

What does "Operationalizing NSX" mean

Operationalizing NSX refers to what happens after the design and implementation of NSX as a software defined networking and security infrastructure. The term "day 2 operations" is often used to refer to what happens after design and implementation of NSX, but to best leverage NSX's capabilities it is important to think beyond just day 2 operations.

How best to optimize NSX usage? Is the organization aligned to take full advantage of what NSX provides? What about IT's operational processes? How might they impact the benefits provided by NSX's software-defined nature? Could they be optimized to realize the full benefits of NSX? How are new users or new applications brought on board? Are there additional considerations due to NSX? How do developers consume NSX? What are the options? The answers to these questions are central to the concept of operationalizing NSX.

Why IT should adjust the way it operates for SDDC/NSX

As pioneer and leading innovator in "software-defined," VMware is uniquely qualified to help an organization's digital transformation and enable their digital business. To be successful with digital transformation, an organization must shift to a service mindset, aligning business priorities and IT imperatives. IT must also transform to truly deliver on this vision. According to a CIO paper ("How IT Organizations Can Achieve Relevance in the Age of Cloud," 2013), "Instead of cost centers that provide capabilities, IT organizations must become internal service providers supplying business-enabling solutions that drive innovation and deliver value... true business partners rather than increasingly irrelevant, cost-centric technology suppliers." To accomplish the shift to become a service provider and ultimately a true business partner, IT must consistently and continuously deliver value to the business at the speed it requires. VMware's software-defined approach provides IT the technical capabilities to enable this agility and speed. Implementing these enabling technologies provides IT the opportunity to change the way it works. Leveraging NSX and SDDC capabilities unlocks IT's ability to quickly respond to changing business needs and provide increased business value.

As a result of working directly with NSX customers, VMware has found that changing the way IT works should be an evolutionary journey across three dimensions - people, process, and tooling - as shown in Figure 1.1.

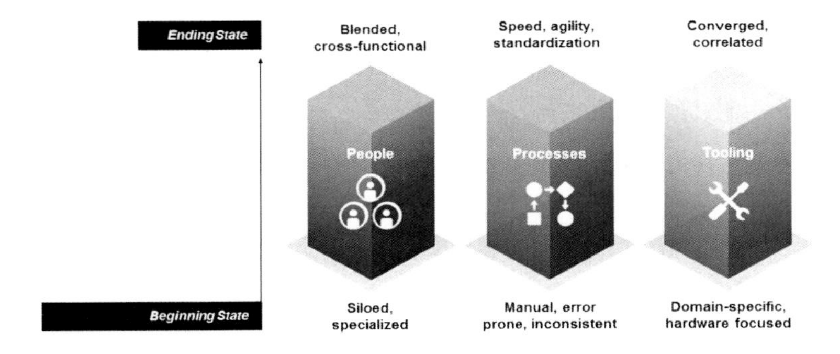

Figure 1.1 Typical beginning state and target end state

Organizations using a traditional IT operating model of siloed functional teams often struggle with SDDC transition. Structured manual processes and domain-specific tooling can hinder NSX operationalization. It is important to work across these three dimensions: breaking down the siloed teams, automating processes to take advantage of software defined capabilities, and moving towards stack aware tooling. This will not only deliver the desired speed and agility, but also help realize the full benefits NSX. This change will not happen overnight. Transforming people, process, and tooling to optimize operation of a software-defined environment should be evolutionary, not revolutionary.

Who the guide is for

Overall this guide is meant both for people new to NSX as well as those who have implemented NSX and are looking to get more out of it. It is intended for:

- Organization level decision makers who are concerned with the people perspective: organizational impacts, team structure, roles and skillsets, culture and mindset
- Managers responsible for or who can influence operational processes such as change and configuration management
- Engineers and technical decision makers responsible for optimizing tooling to be more effective in a software defined environment

In short, it provides a little bit of something for everyone who should be involved in operationalizing NSX.

What it will teach

This goal of this guide is to present what is required to operationalize an NSX environment from an introductory perspective. References to more detailed documentation will be provided where applicable.

Why it matters

For companies who have invested in NSX and the broader SDDC suite, the technology enables them to provide greater value to developers, application owners, and even end users. It is in their best interest to derive the greatest value from this investment. The guidance provided in this book will help individuals understand how best to unlock the business value of NSX and SDDC. Software-defined solutions are the future; this book can help individuals build onto their career skillset while increasing their immediate personal value to their employer.

How to proceed

Different sections are more valuable to different readers. Organization level decision makers would benefit from the entire book for context and increased understanding of how to unlock NSX's potential. "Measuring Results" and "People Considerations" are of primary importance, followed by "Intelligent Operations." Managers interested in optimizing operational processes should focus on the process related sections in "Intelligent Operations." Technically-inclined individuals can focus in on the monitoring and troubleshooting sections of "Intelligent Operations."

Chapter 2

Measuring Progress

It is important to make some changes to fully realize the benefits of NSX; attempting to fit NSX into existing models is not a path to optimal value realization. As with any change, especially transformative change, it is valuable to be able to track progress and validate results.

Why measure

What good is undertaking the effort to change without measure to prove demonstrable progress and return on investment? It is more powerful to back up talk of success with data - detailing the operational benefits and business impact associated with reduced application deployment time - rather than simply stating things are faster.

It is also good to measure progress and results to drive conversations that help teams improve. These efforts should focus on providing positive reinforcement to encourage continuous improvement rather than merely evaluating team performance.

How and what to measure

Begin with a good understanding of the current state of operational performance. Establish a baseline to measure improvement against, tracking progress to ultimately claim success. This book provides suggested key performance indicators to use as a starting point for measurement in specific, important areas.

Finding quick wins and with whom to share results

Quick wins are key. Do not try to boil the ocean by making many changes at once. Start by focusing on processes that will deliver the most value to IT or end users with the least amount of effort. Choose a specific application or service for initial focus and create a tiger team as discussed in the "People Considerations" section.

For example, create a load balancer and apply it to the front end of a specific application. How long would it take to perform those steps in the physical world? How long would it take to purchase, burn-in, configure, and deploy a load balancer for a new application? Compare that to defining and deploying a software-based load balancer in NSX. What were the savings - in time and money - in pure deployment time as well as cost of both people and licensing? An examination such as this will help make the concept real.

With whom should results be shared? Up the management chain, down the management chain, and laterally to other teams within IT! This is a new technology and a new way of working with networking and security. Reinforce up the IT management chain how virtualized network and security implementation produces valuable IT and business outcomes. Further incentivize success through acknowledgement and recognition down the management chain. Market success laterally across IT to both quiet the naysayers and increase involvement as application of the virtualized network and security expands. Finally, do not forget about business stakeholders. Be sure to actively communicate and market success to them if it provides a demonstrable business outcome to which they can relate.

Focus on quick wins and share the results up the management chain, down the management chain, and laterally to other teams within IT.

Chapter 3

People Considerations

Whether a smaller IT organization just getting started or a larger IT organization focusing on truly becoming a service provider to business stakeholders, the focus is ultimately on delivering business value with software defined networking and security. NSX is an enabling technology but to fully realize its potential for contributing to business VMware recommends some operating model optimizations. The starting point is the people perspective. Aligning team structures, roles, and skillsets along with affecting cultural and mindset changes are the basis for these operating model optimizations.

Why SDDC (and NSX) changes the people equation

Operationally, IT organizations are traditionally composed of technically-aligned functional groups. To date, they have been able to "get by" even with a significant investment in compute and memory virtualization-based infrastructure environments. These organizations have gained efficiencies and seen improvements in virtual workload deployment times by automating components of their virtual workload deployment process, but these are at best incremental. Regardless of the activities in a virtualized environment, if multiple functional teams are needed to complete a task like deploying a new application or service, it is not an organization that can quickly respond to business changes and requirements.

The challenge does not stop with technically-aligned functional teams. IT organizations of any size are inevitably organized in plan, build, and run silos. Architects make technical decisions and design solutions; engineering teams build and test the solutions architects provide them; and the solution is handed to IT operations to run. How long does this end-to-end process take? Or worse, how prepared is IT operations to run it?

This can extend beyond putting full solutions in production which can be still more problematic. One customer applied a subset of this process to developing new and modifying existing vRealize® Operations Manager™ dashboards. Operations had to provide requirements to a tools engineering team who provided the dashboard two months later. Once the operations team received the dashboard, it no longer provided what they needed.

By providing in software what was once only hardware-based, NSX and SDDC enable fully virtualized compute, memory, network, and security infrastructure solutions. This allows the workloads running on them to be quickly implemented and changed. With the additional automation opportunities, IT has the potential to provide previously unheard of levels of business value. This potential will never be reached using existing IT functional grouping constructs. Maintaining these constructs will also prevent IT from adopting the Agile methodologies being leveraged in application development; methodologies that lend themselves to making IT more responsive to and faster in delivering solutions that meet changing business needs.

Break down silos across two dimensions to be successful: plan-build-run and technically-aligned functional teams.

Team construction

VMware recommends a new team structure to leverage NSX capabilities in the context of a Software Defined Datacenter. The goal is a blended or integrated team; a team consisting not only of cross-functional technical skills but cross-domain roles. This is aimed at creating a much closer relationship between architecture, engineering, and operations for an SDDC-based environment including NSX. This team will serve as the focal point for all decisions and actions regarding the environment.

The goal in creating a blended team is to break down the cross-functional and cross-domain IT silos, as shown in Figure 3.1, that inhibit agility and execution speed. It intends to replace them with a team built for tight collaboration and focus. Creating such a team results in faster and better decision making, reduced time to problem resolution, and more operationally ready solutions.

Figure 3.1 Cross-domain and cross-functional silos

Why is creating a blended, cross-domain team important for success with NSX and SDDC? Organizations can no longer afford to have distinct plan, build, and run teams if they want to realize the agility afforded by software defined infrastructure. Operations needs to have more involvement in architecture and design decisions; for example, will the resulting infrastructure be built with rapid change in mind? Do the architects and engineers really appreciate the operational flexibility of a software defined infrastructure? Architects and engineering functions should wear the operational pager once a month to gain that appreciation. A tight feedback loop is needed between plan, build, and run when working with a software defined infrastructure. This only works well, especially at scale, when these cross-domain functions are part of a blended team.

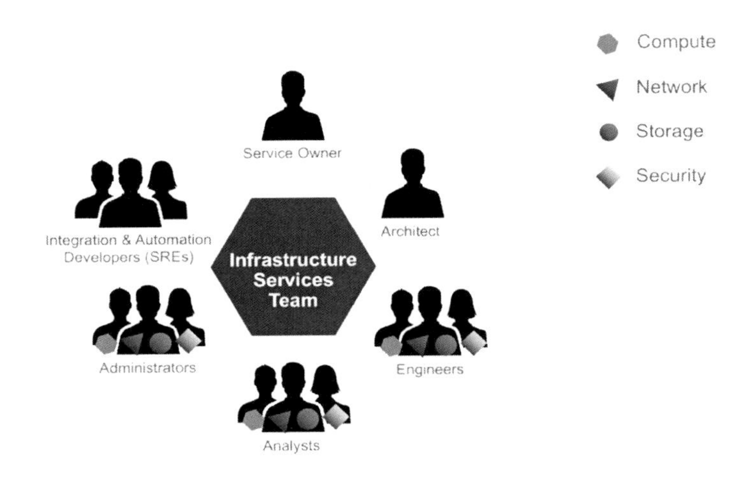

Figure 3.2 Blended team

The same is true for breaking down technical silos and creating a blended, cross-functional team. This concept is shown in Figure 3.2. Relying on hand-offs between systems, network, security, and storage to accomplish a result in a software defined infrastructure is antithetical to achieving agility and speed of execution. Where a line of business wants to rapidly begin developing a new mobile application to address customer feedback, waiting for IT to deliver infrastructure does not a happy customer make. Functional team ticket system- and backlog-driven tasking does not lend itself to taking advantage of NSX capabilities. These must become a thing of the past to be successful in a software defined data center.

This blended team can be physical or virtual; matrixed or under a single manager. There is no best answer as these decisions tend to be very company-specific, often involving politics and cultural dynamics. Achieving the guiding principle of creating a blended, cross functional, cross-domain team is what matters, and it can work in various organizational and reporting structures. The common success factors are: the team is built with a clear purpose and shared objectives; they are incentivized on achieving team objectives more so than individual objectives; and they are as self-sufficient as possible with decision authority and the associated accountability to achieve their objectives. In addition to technical and domain skills, it is important to have an enthusiastic team of change agents who can naturally act as evangelists. This is a new technology and way of working; they must embrace this change and make it infectious to others. This is essential to ensure on-going success as software defined networking and security start to represent a larger percentage of the overall infrastructure and become more business critical.

Forming a blended, cross-functional, cross-domain team is paramount; whether it is virtual, physical, matrixed, or under a single manager is a company-specific organizational decision.

Creating such a team as well as its size is circumstance dependent. Is IT just getting started with a new NSX implementation? Is it adding NSX to an existing environment or is this part of a larger, greenfield software defined data center implementation? Are they experimenting with software defined networking and security or is this an all-in situation where the company has made a strategic decision to go with a software defined data center?

For organizations just getting started with a new NSX implementation or experimenting with software defined networking and security, the process shown in Figure 3.3 may work well.

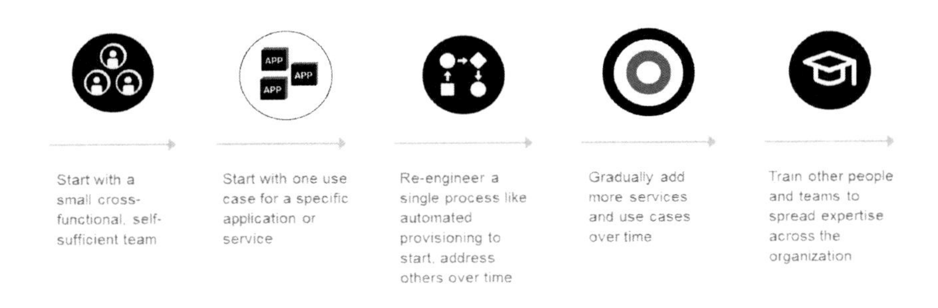

Figure 3.3 Getting started

Experience shows that initially taking a tiger team approach to work best. Create an initiative focused around some aspect of IT automation for example. A common example is automatically provisioning virtual networking and security for three-tier applications using VMware vRealize® Automation™. To do this, define and create a blueprint that can be standardized and used repeatedly. This type of effort requires input from the application team and all the virtual infrastructure teams: compute, storage, network, and security. To truly control the resulting automatic provisioning service end-to-end - plan, build, and run - it requires involvement from architecture, engineering, and operations. Pull together a cross-functional, cross-domain tiger team, to drive this consensus building initiative. Remember to baseline current and resulting metrics around provisioning network and security for these three tier application infrastructures - both to prove the benefits of a blended team as well as to put forth a data-driven argument for making it the standard going forward. Keep the decision-makers informed from the beginning of the initiative, provide them with regular updates, and schedule a final read-out session to sell the value of a blended team based on the data-driven, business stakeholder impacting results.

Start with a small, blended tiger team focused on a single use case for a specific application or service, then expand over time.

The approach for constructing a blended team in the context of a more strategic software defined data center initiative depends on whether or not the need for operating model changes are accepted. If the need is accepted and already part of the strategic plan, the question often comes down to lines of delineation. What does the existing networking team continue to support versus the team responsible for the virtualized software defined data center infrastructure? What about security? A useful delineation can be the physical firewall and layer 2. Everything on the SDDC side of the firewall above layer 1 should be owned by the team responsible for the software defined data center.

A customer in the US has created a blended team including network and security engineers. Their legacy networking team retains responsibility for all physical networking while the network engineers on the blended team have responsibility for software defined networking within their SDDC environment. Security engineers on the blended team are responsible for all east-west security within the SDDC while their IT security team and Security Operations Center (SOC) is responsible for all north-south security at the SDDC boundary. They had some early reticence to include security engineers in the blended team but they are now happy they did so. This alignment simplified integration of security thinking into the architecture, design, and operations of the software defined data center rather than relying on external auditing and remediating security.

The next chapter details the skills necessary to affect this, but from a functional perspective, compute, network, storage, and security should all be represented on the team responsible for the SDDC. The bigger challenge may be getting security representation on the team. This is often the more controversial of the two when creating a blended team with all the functional skills represented. The argument is for general IT security policy creation, monitoring, auditing, and enforcement to remain with the IT InfoSec team while the NSX-based implementation, monitoring, and remediation of IT security policy should reside on the blended team for operational agility, efficiency, and speed of execution.

Including a security role for NSX in a blended team may seem less obvious, but its absence will inhibit success.

If operating model optimization is not an inherent component of a larger NSX or SDDC initiative, fall back on the same approach described to simply get started with NSX. Create a tiger team focused around a single activity or process that addresses the business justification. Create a baseline, track metrics, then sell the results in the context of demonstrable IT and business outcomes.

Creating a blended team is critical to fully leveraging NSX capabilities and achieving game-changing success in the software defined data center. Without the commitment to create a blended team, expectations of dramatic improvements in agility or speed of execution will need to be managed. In this case it is still recommended to move forward with a tiger team exercise while measuring results against a baseline. Use this to champion additional temporary tiger teams to drive business impacting improvements using software defined network and security capabilities. Following this approach can provide incremental improvements, but the challenge will be sustaining the improvement operationally. Continue toward the ultimate goal of gaining mindshare for creating a blended team.

Roles & skillsets

This section provides guidance on the recommended roles for software defined networking and security. It also describes key responsibilities, skillsets, and education associated with each role. This section only addresses the software defined networking and security roles, not all the roles recommended for a fully blended team. For more information regarding the other roles in a blended team please download the "Organizing for the Cloud" whitepaper referenced in the "Where to go for more information" section.

It is also important to remember that this is a discussion of roles, not headcount. In small and mid-size environments, roles may be combined into a smaller number of individuals. In large, business critical environments, there may be multiple individuals providing a single role perhaps with even greater specialization.

The recommended approach to a fully blended team in the broader software defined data center context tends towards full stack knowledge with some specialization. Though it addresses the roles only in the context of software defined networking and security, this section does provide a description of how the role is impacted in an approach to fully blended teams for the general SDDC.

While describing both a network and security architect role, an overall Cloud Architect role is recommended for a general SDDC blended team. This role would have more general network and security skillsets for a software defined data center blended team. In that case, focus the network and security-specific skills respectively in network and security engineers. They will work closely with the cloud architect, providing the needed network and security subject matter expertise.

Table 3.1 Architecture roles

Role	Responsibilities	Skillsets and Education

Cloud Network Architect	• Identify and prioritize use cases and business requirements to address with virtualized networking and security
	• Design logical network services for availability, capacity, mobility, recoverability, and data protection
	• Design standards and templates for automated virtualized networking provisioning and configuration management
	• Verify virtualized network solutions by developing and validating tests to ensure the success of addressing use cases and requirements
	• Identify modern tools for virtualized network orchestration and automation, and day-2 operations (e.g., visibility, monitoring, troubleshooting)
	• Guide the virtualized networking implementation strategy and assist operationally with on-boarding new applications and services; establish new, optimized processes
	• Provide level 3 support as needed to work within defined SLA or OLA resolution period
	• Assist in defining and evolving overall IT network architecture and standards that maximize the synergy, reuse, and value of NSX over time.
		• Cross-domain skills (e.g., virtualized network & security, VMware vSphere®, virtual distributed switching, network protocols)
		• Education
		  ◦ Data Center Virtualization Fundamentals
		  ◦ NSX Install, Configure, & Manage or NSX for Internetworking Experts Fast Track
		  ◦ NSX: Design & Deploy; NSX: Troubleshooting and Operations
		  ◦ NSX Hands-on Labs
		• Certification: VCDX-NV

Role	Responsibilities	Skillsets and Education
Cloud Security Architect	• Identify and prioritize use cases and business requirements to address with virtualized security • Determine technical security requirements and translate them into security policies and standards; plan and guide the implementation of these security controls and solutions • Design standards and templates for automated virtualized security provisioning and configuration management • Verify virtualized security solutions; develop and implement efficient validation controls and tests • Determine auditing and reporting processes for virtualized security impacting compliance • Provide level 3 support as needed to work within defined SLA or OLA resolution period • Conduct security risk assessments for cloud workloads and infrastructure; provide authoritative advice and guidance on security strategies to manage the identified risk	• Cross-domain skills (e.g., virtualized network & security, vSphere, virtual distributed switching, access control) • Education ° Data Center Virtualization Fundamentals ° NSX Install, Configure, & Manage or NSX for Internetworking Experts Fast Track ° NSX: Troubleshooting and Operations ° NSX security-related Hands-on Labs

The Network Engineer and Security Engineer roles are key for blended teams, whether they be specifically NSX-focused or broader SDDC-focused. These network and security specializations are focused on the engineer roles regardless of whether they are providing deeper network and security subject matter expertise to the architect roles or designing profiles and policies implemented by the administrator roles.

The Cloud Automation & Integration Developer role is also included in Table 3.2 as it is a critical role in the modern software defined data center. This role is important regardless of context – either in an NSX-focused a SDDC-focused blended team. Automation is absolutely key to success going forward. Automation-related workflow development must move beyond administrators writing scripts and toward formal software development. This is a focus of the Cloud Automation & Integration Developer.

Table 3.2 Engineering roles

Role	Responsibilities	Skillsets and Education
Cloud Network Engineer	• Low-level design, deployment, and testing of the virtualized network functions that realize the virtualized network service; definition of the virtualized network function configurations; validation of virtualized network services functionality; operationalizing virtualized network services • Supports the Cloud Network Architect role in designing cloud network services; translating the requirements into blueprints and configuration templates for the network functions • Ensure fulfillment of requirements – including capacity, availability, security, compliance, and SLAs • Deploy, test, validate, and manage monitoring/ troubleshooting tools, processes, dashboards, runbooks • Work with the Cloud Automation & Integration Developer role to design, develop, test and deploy custom workflows and scripts within the virtualized network infrastructure for use with integration, orchestration, deployment, monitoring, compliance, or other routine tasks. • Provide level 3 support as needed to work within defined SLA or OLA resolution period • Diagnose and analyze root cause of issues; apply patches and fixes as needed • Implement routine, approved, and exception changes in the infrastructure • Assess and test upgrades and patches for virtualized networking and security infrastructure and tools	• Cross-domain skills (e.g., virtualized network & security, vSphere, virtual distributed switching, network protocols) • Education ° Data Center Virtualization Fundamentals ° NSX Install, Configure, & Manage or NSX for Internetworking Experts Fast Track) ° NSX: Design & Deploy ° NSX: Troubleshooting and Operations ° NSX, vRealize® Operations™, vRealize® Log Insight™, & vRealize® Network Insight™ Hands-on Labs • Certification: VCP-NV, VCIX-NV

Role	Responsibilities	Skillsets and Education

Cloud Security Engineer	• Translates IT security policies into security controls appropriate to SDDC-based cloud environment
	• Designs, implements, deploys, configures, and monitors the security solutions and procedures for the SDDC-based cloud environment
	• Assists the Cloud Security Architect role in designing and planning the cloud security architecture, security policies, and security processes.
	• Works with the Cloud Automation & Integration Developer role to develop the workflows that orchestrate the security controls according to the security policy; develop security monitoring and remediation solutions, workflows and integrations.
	• Investigate identified security breaches in accordance with established procedures; recommend and implement any required action
	• Work with the IT security functional team to ensure that cloud security services integrate with existing tools and processes; validate that these fulfil IT security & compliance requirements
	• Manage security information – including logging, auditing, and reporting capabilities
	• Diagnose and analyze root cause of security-related issues; apply patches and fixes as needed
	• Implement routine, approved, and exception security-related changes in the virtualized infrastructure
	• Assess and test upgrades and patches for virtualized networking and security infrastructure and tools
		• Cross-domain skills (i.e., virtualized security, vSphere, access control)
		• Education
		  ◦ Data Center Virtualization Fundamentals
		  ◦ NSX Install, Configure, & Manage or NSX for Internetworking Experts Fast Track
		  ◦ NSX: Troubleshooting and Operations
		  ◦ NSX (security-related), vRealize Operations, vRealize Log Insight, & vRealize Network Insight Hands-on Labs
		• Certification: VCP-NV

Role	Responsibilities	Skillsets and Education

Cloud Automation & Integration Developer	• Work with the Cloud Network and Cloud Security Engineers to design and develop code to enable integration with other systems or tools
	• Work with the Cloud Network and Cloud Security Engineers to establish integration and automation monitoring
	• Work with the Cloud Network and Cloud Security Engineers to establish automated virtualized networking and security service provisioning; establish event and incident remediation wherever possible and appropriate
		• Skills: PowerShell, Python, Ansible), Configuration Management tools (Chef, Puppet), Orchestration tools (VMware vRealize® Orchestrator™), NSX API, VMware vRealize Automation® API, VMware vRealize Operations API
		• Education
		  ◦ Data Center Virtualization Fundamentals
		  ◦ Data Center Automation with vRealize Orchestrator and PowerCLI
		  ◦ VMware Cloud Orchestration and Extensibility
		  ◦ vCenter Orchestrator: Develop Workflows

The final two roles are the Network Administrator and Security Administrator. In addition to day to operations (e.g., backup & restore, upgrade & patching), the administrator roles are heavily focused on proactively monitoring and remediation of the virtualized network and security infrastructure. They are also responsible for working with the engineer and developer roles to customize the monitoring tools, continuously improve their proactive and predictive capabilities. The goal is to minimize the actual number of incident tickets received by identifying and remediating issues before they become service or application disrupting.

Table 3.3 Administrator roles

Roles	Responsibilities	Skillsets and Education

Cloud Network Administrator	• Monitor physical and logical network infrastructure and act on events before they affect services
	• Proactively monitor network performance (e.g., latency, throughput), health (e.g., faults, failures, connectivity), availability, and configurations
	• Update and maintain virtualized networking infrastructure by utilizing alarm/alert mechanisms
	• Provide level 3 support for virtualized networking; narrow down the problem in physical or logical using modern tools
	• Investigate and diagnose logical network infrastructure and services incidents
	• Ensure solutions and fixes are applied to recover from network incidents
	• Implement and apply virtualized network policies designed and test by the Cloud Network Engineer
	• Backup and restore of NSX Manager data (e.g., system configuration, events, audit log tables)
	• Upgrade and patch virtualized networking and security infrastructure and tools
	• Define, develop, and test custom dashboards, super metrics, reports, etc., for virtualized networking infrastructure and services; work with the Cloud Automation & Integration Developer role to implement remediation automation capabilities
		• Cross-domain skills (i.e., virtualized networking, vSphere, virtual distributed switching, network protocols)
		• Education
		  ◦ Data Center Virtualization Fundamentals
		  ◦ NSX: Troubleshooting and Operations
		  ◦ vRealize Operations for Operators
		  ◦ NSX, vRealize Operations, vRealize Log Insight, & vRealize Network Insight Hands-on Labs
		• Certification: VCP-NV

Roles	Responsibilities	Skillsets and Education

Cloud Security Administrator	• Monitor virtualized security services and act on events before they affect services
	• Proactively monitor virtualized security services performance, health, availability, and configurations
	• Engage in escalations affecting security in the SDDC-based cloud environment
	• Investigate and diagnose security incidents in the SDDC-based cloud environment
	• Ensure solutions and fixes are applied to recover from security incidents in the SDDC-based cloud environment
	• Implement and apply virtualized security policies designed and tested by the Cloud Security Engineer
	• Understand, apply, and maintain specific security controls in the SDDC-based cloud environment as required by corporate security and compliance policies
	• Assist in the performance of SDDC-based cloud environment audits
	• Ensure SDDC-based cloud workload and infrastructure comply with organizational standards for logging – including content, format, and location.
		• Cross-domain skills (i.e., virtualized security, vSphere, access control)
		• Education
		  ◦ Data Center Virtualization Fundamentals
		  ◦ NSX: Troubleshooting and Operations
		  ◦ vRealize Operations for Operators
		  ◦ NSX (security-related), vRealize Operations, vRealize Log Insight, & vRealize Network Insight Hands-on Labs
		• Certification: VCP-NV

This section described roles, not headcount or individuals. Depending on the scale and business criticality of an environment, multiple roles may be filled by a single individual or multiple individuals may fill a single role. The most important point is that these roles and skillsets exist in the blended team.

Focus on full-stack knowledge with some specialization when filling roles and putting together training plans for blended teams.

Culture & mindset

Culture and mindset is the equivalent of the organization's DNA – the values and beliefs that shape how people behave and create the organization's culture. This is the single most impactful factor influencing success or failure for adoption of software defined networking and security. It is also the most difficult to change. This is one of the first things to assess when getting serious about implementing NSX.

The ideal culture is one that embodies collaboration and is guided by a focus on business outcomes. It consists of a team of full stack generalists with some specialization who are interested in continuous learning and improvement, a team both responsible and accountable for achieving objectives they acknowledge as owning. Many of these characteristics are associated with the Agile culture popular in application development teams. This represents a rare breed of organization, though it is critical for success.

How can an organization move toward this state? There are entire books devoted to answering this question, but some guidance includes:

- Start with leadership. Cultural change must be embraced from the top down, even if it is only realized in the NSX or SDDC blended team.
- Leadership must articulate a clearly defined direction for the blended team and modify incentives to reinforce shared team objectives
- The team must be given ownership, responsibility, and accountability for achieving the stated purpose and shared objectives
- The members initially selected for the blended team must be like-minded change agents; they must be open-minded and passionate about instituting the culture
- Team successes must be recognized and advertised; the goal is to make their behavior aspirational for others in IT. What the team is doing and how they are progressing should be actively marketed within IT to key business stakeholders

Overcoming cultural challenges is the biggest hurdle and should be explicitly addressed from the beginning.

KPIs to measure progress

Key Performance Indicators (KPIs) are used to measure progress towards target state objectives. These objectives, in turn, should clearly communicate the definition of success. Target state objectives are company or organization specific, with their goals tied to distinct IT and business outcomes. Table 3.4 offers some general purpose KPIs for measuring progress towards establishing software defined networking and security blended teams. As establishing baselines for comparative measurement further increases value, build these into the KPIs from the beginning.

Table 3.4 Sample team structure KPIs

Objective	KPI	
Self-sufficient team	• % of escalations resolved within the blended team • Average time to resolve escalations with the blended team	These KPIs provide a measure of blended team efficiency in resolving issues versus the baseline of how long it previously took to resolve escalations across siloed teams.
Shared objectives	• Team-based annual review criteria as a % of team member's review criteria • Average time to complete cross-functional activities within the blended team	These KPIs provide a measure of a blended team's efficiency in completing cross functional activities (e.g., on-boarding a new application) when they have team-based objectives versus the baseline of completing a similar activity involving siloed teams with competing objectives.

Chapter 4

Process Considerations

Intelligent Operations

Most IT organizations cannot break out of firefighting mode and are constantly reacting to monitoring events and alerts generated in their environment. They are also plagued with laborious operational tasks consisting of error-prone manual steps. Applying this mode of operation to a NSX-based environment will not deliver the full benefit of software defined networking and security. It will minimize the agility and speed of execution opportunities provided by NSX software defined networking capabilities. The concept of "Intelligent Operations" refers to a modern, proactive mode of operation that optimizes and automates processes and workflows to take advantage of software defined networking and security capabilities. This is contrasted against a more traditional reactive mode of operation with its constant focus on break fix activities and lack of time for innovation and improvement. It is also about using the right tools for the job - tools purpose built for Intelligent Operations in a software-defined infrastructure.

This chapter outlines intelligent operations as it relates to software defined networking and security, focusing on optimizing and automating the activities and processes most impacted by NSX. Examples will be based on VMware's vRealize® intelligent operations tools as well as NSX open source tools. Descriptions of general tool capabilities will be used to aid in substitution as necessary.

Proactive monitoring for performance and availability

Proactive monitoring is more about mindset change than technology. It involves proactively monitoring key metrics, analyzing potential issues, and remediating those issues before they become service, application, or end-user impacting. Even when focused on software defined networking and security, proactive performance and availability monitoring puts more emphasis on monitoring from a service- or application-centric perspective. An example may be Infrastructure as a Service (IaaS) or Platform as a Service (PaaS) provided to application developers or Data Analytics as a Service consumed by lines of business. While implementing proactive monitoring depends on a mindset shift, the selection of proper monitoring tools is also an important factor in its overall effectiveness.

What does it mean to put more emphasis on monitoring from a service- or application-centric perspective? First and foremost, this means monitoring the full, integrated stack where an application or service is running. This involves monitoring the VMs that comprise the application or service, the hypervisor and host, storage, and the end-to-end network path - anything that impacts the application or service if there is an issue in the stack. It includes monitoring common metrics such as CPU performance, memory utilization, network throughput, and storage throughput latency. This is all done within the context of any service level agreements associated with the application or service. If a service level agreement is involved, best practice includes setting monitoring thresholds at some percentage below the service level threshold to provide a buffer, allowing time to troubleshoot before an issue impacts the application or service.

It is worth adding a quick check of end-to-end network health for tier one applications to the daily routine. This is easily done with vRealize Network Insight's topology-based 360° visibility and analytics, as shown in Figure 4.1

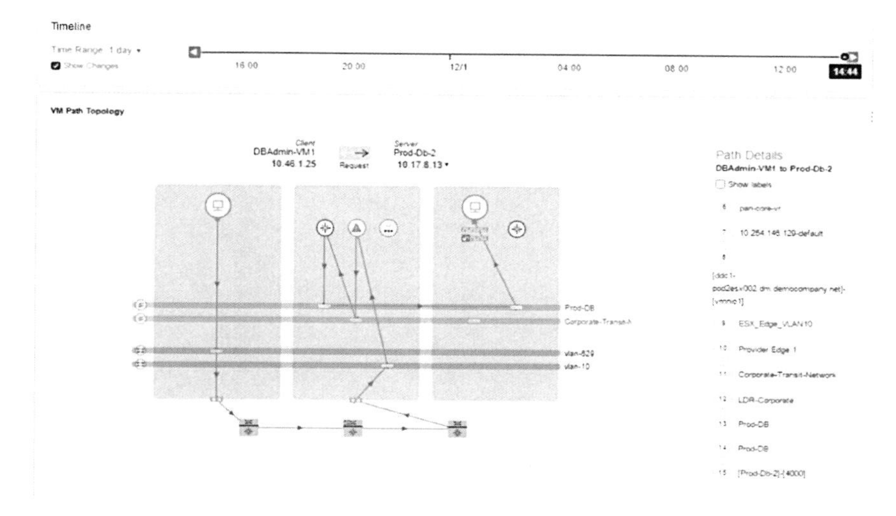

Figure 4.1 Example of vRealize Network Insight's 360° visibility

What if the NSX Edge device providing the load balancing or VPN service supporting an application starts trending towards lower throughput or high latency? One effective tool to monitor the NSX Edge device is vRealize® Operations Manager™. vRealize Operations Manager can monitor the NSX Edge as a virtual machine, taking advantage of vRealize Operations Manager's intelligent analytics engine. The intelligent analytics engine can learn the normal behavior of the NSX Edge VM, then alert when trends outside of its learned behavior or exhibits anomalous behavior.

While monitoring will identify issues with the NSX Edge VM, it may not be straightforward to link the problem back to the application in the environment. Use of vRealize Network Insight to review the status of all application components in the full end-to-end path will greatly aid in proactively managing application performance and availability.

KPIs

Table 4.1 Sample proactive performance and availability monitoring KPIs

Objective	KPI	
Proactive performance monitoring	• Number of performance-related issues detected and resolved before they become service or application impacting incidents • % reduction in performance-related incidents month over month	These KPIs provide a measure of the effectiveness of proactive performance monitoring over time.
Proactive availability monitoring	• Number of availability-related issues detected and resolved before they become service or application impacting incidents • % reduction in availability -related incidents month over month	These KPIs provide a measure of the effectiveness of proactive availability monitoring over time.

Proactive capacity monitoring and planning

Similar to performance and availability monitoring, proactive capacity monitoring and planning involves shifting a mindset to proactively identifying and remediating capacity issues before they become service, application, or end-user impacting. Unlike proactive performance and availability monitoring - which is best approached from a service- and application-centric perspective - proactive capacity monitoring and planning for software defined networking and security focuses on the NSX components themselves. There are three main areas of focus for capacity monitoring and planning of a software defined network and security-based environment: NSX Manager/ Controllers, NSX Edge devices, and workload clusters utilizing NSX.

It is important to always monitor NSX Manager and NSX Controllers from a capacity perspective, especially so in a multi-site environment. When running a multi-site NSX Manager configuration, be certain to monitor inter-site network-related capacity. NSX Managers need to synch with one another in a multi-site NSX Manager configuration, and this ability is directly related to inter-site throughput and latency. There is a maximum supported latency of 150 milliseconds between sites for synchronization.

In a multi-site environment, NSX Controllers interact with remote vSphere hosts. As with NSX Managers, the NSX Controller interaction with remote vSphere hosts in multi-site environment is also dependent on inter-site network throughput and latency. NSX Controllers are also sensitive to disk latency, which should be monitoring in both single and multi-site environments.

NSX Edge devices are one of the most important components to proactively monitor. NSX Edge device services (e.g., load balancing, VPN) have a direct impact on services, applications, and end-users. CPU, memory, and throughput capacity can directly impact load balancer and VPN effectiveness. Valuable metrics to monitor include: CPU Demand, CPU Run Queue, CPU Swap Wait, CPU I/O Wait, Ram Free, Ram Committed, Page-in Rate, and Network Usage. It is also recommended to monitor the number of NSX Edge devices per host, ensuring the maximum recommended number is not exceeded.

It is useful to proactively monitor distributed firewall memory usage at the host level in workload clusters as firewall rules are replicated across vNICs. If the memory usage maximum is reached, no additional firewall rules can be deployed.

KPIs

Table 4.2 Sample proactive capacity monitoring and planning KPIs

Objective	KPI	
Proactive capacity monitoring	• Number of capacity-related issues detected and resolved before they become service or application impacting incidents • % reduction in reported capacity-related incidents month over month	These KPIs provide a measure of proactive capacity monitoring effectiveness for example of the Edge cluster. (Requires Proactive Issue Resolution checkbox along with Capacity Incident Category in ITSM tool)
Proactive capacity planning	• Average amount of time in advance of on-boarding a new application or service that requirements for additional NSX Edge cluster capacity are identified • Average amount of unused capacity in the NSX Edge cluster month over month	These KPIs provide a measure of the effectiveness of proactively planning capacity needs for new applications as well as the accuracy of proactive capacity planning.

Change management

Software defined networking and security can and will have an impact on change management. NSX is software-based, leading to fewer change management activities in the physical network. This can reduce the overall change scope, lessening the impact on infrastructure and dependent applications and services. Since NSX is policy-based, it can be automated to avoid the most significant cause of change back-outs (i.e., manually applied changes). As NSX is software defined and policy-based, it is easier to perform validation tests in advance of a change being applied in production.

This is especially true when automatically deploying NSX capabilities via tools with blueprint capabilities (e.g., vRealize Automation). In this model, blueprint changes can be handled through the change control process while workload or application deployment into production is a pre-approved change. This is most applicable where the same blueprint is used repeatedly for deploying workloads or applications into production.

Another example of change management process modification is the use of NSX for micro-segmentation. In this example, a firewall policy model for micro-segmentation consists of five types of firewall rules:

- Emergency firewall rules used for quarantine and/or allow rules for example
- Infrastructure firewall rules - global firewall rules applied to common object or services (e.g., AD, DNS, NTP, DHCP, management servers)
- Environment firewall rules, including firewall rules between zones (e.g., production vs development), PCI vs non-PCI, and inter-business unit rules
- Inter-application firewall rules, rules between applications
- Intra-application firewall rules (e.g., rules between application tiers, rules between micro-services)

Based on this example, how might change management be applied in the micro-segmentation scenario shown in Figure 4.2 where each bubble represents an application isolated using security policies?

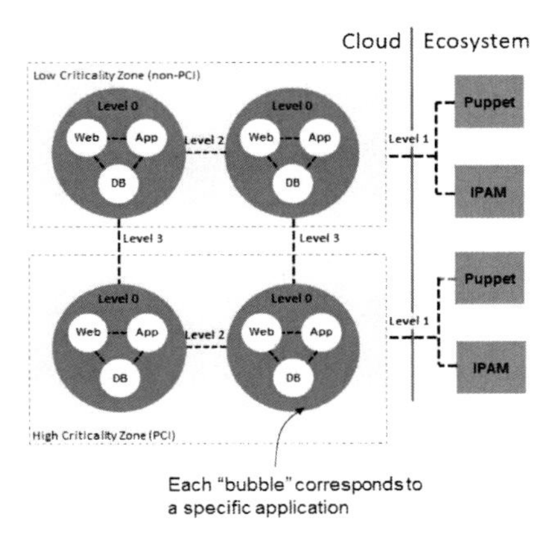

Figure 4.2 Example of micro-secgmentation in a multi-application deployment

The following bullets discuss the considerations and impact on the standard change management process for this scenario:

- Level 0 represents isolated, multi-tier applications. Intra-application firewall rules are controlled by application developers and may not require any change approval.
- Level 1 represents infrastructure services defined in the deployment blueprint and accessed at provisioning time. These are controlled and validated by the team providing the IaaS workload. Change management is applied to the blueprint, so deployment from the blueprint is pre-approved.
- Level 2 represents inter-application firewall rules allowing communication between two applications within a low criticality zone. Applying these firewall rules could be pre-approved since they can be extensively validated prior to production deployment. Though pre-approved, automatically creating, filling out, and closing a change ticket when the firewall rules are applied could be done for auditing purposes
- Level 3 represents inter-application firewall rules similar to level 2, but controls communication between applications in different criticality or confidentiality zones. Applying these firewall rules should be subject to standard change control procedures.

The potential impact on change management can have a direct effect on the agility and time to value business stakeholders will experience.

Because of the potential impact, it is a best practice to actively monitor the effect of the normal changes as they are being made in production. This also represents another advantage of a blended team model; all team members will be involved by default. They will know to monitor normal changes as they are being made and will be extra vigilant with monitoring when pre-approved changes take place, greatly enhancing the likelihood of successful change management.

KPIs

Table 4.3 Sample change management optimization KPIs

Objective	KPI	
Positive impact on change management by having a blended team	• Ratio of terminated changes to successful changes	This KPI provides a measure of blended team efficiency in planning and executing changes versus the baseline of the same metric for changes involving siloed teams.
Increase in number of automated standard changes	• Total number of automated changes versus manual changes per month	This KPI provides a measure of the number of automated changes completed, indirectly reflecting a reduction in the cost of operations.
Increase number of pre-approved standard changes	• Total number of pre-approved Standard Changes compared to Normal and Emergency Changes	This KPI provides a measure of what should be an upward trend in the number of standard changes with a downward trend in the number of emergency and normal changes.

Configuration management

The software defined nature of NSX simplifies configuration management when coupled with automation. When using vRealize Automation to perform workload deployments, all the software-defined, logical components are tracked in blueprints that can be put under version control. Configuration information can be automatically inserted, updated, and marked as decommissioned or deleted in a CMDB as part of a vRealize Orchestrator workflow invoked from vRealize Automation. Either in conjunction with vRealize Automation blueprints or used stand-alone, Puppet manifests, Chef recipes, or Ansible playbooks can be used to keep NSX-related configurations consistent. Any of these configuration management tools can automatically check for and remediate configuration drift.

An example of using an Ansible playbook to manage NSX logical switch state can be found in the "Automation Leveraging NSX REST API" a link for which is provided in Table 7.1 in the "Where to go for more information" section.

If there is no corporate CMDB or it does not include NSX network or security configuration information, there are still solutions for auditing configuration changes. NSX components can send their logs to remote syslog servers or vRealize Log Insight. Using vRealize Log Insight makes it easy to filter audit log files for configuration change instances. These can then raise security alerts or generate reports. The NSX Controllers also keep track of all deployed software defined networking components. These can be extracted and reported on through the NSX Manager REST API.

KPIs

Table 4.4 Sample configuration management optimization KPI

Objective	KPI	
A change record should exist for any configuration changes	• Number of configuration changes without a corresponding change record	Whether updating a blueprint, Puppet manifest, Chef recipe, Ansible playbook, or creating or changing a configuration item in a CMDB, there should be a record in the change management tool. The change record could have been manually or automatically created.

Provisioning NSX capabilities

Many organizations have focused on automated provisioning of virtual infrastructure and workloads. This may involve offering IaaS to users or simply expediting the deployment of virtual infrastructure and workloads by IT on the user's behalf.

Without the capabilities of NSX, network and security aspects must still be provisioned manually. This slows down the provisioning process, impacts user wait times, and increases the probability of introducing human error. These errors can lead to misconfigurations that result in an unusable environment and further user wait time as the misconfiguration is corrected. The software-defined nature of NSX helps avoid such problems, allowing IT to fully automate the network and security aspects of virtual infrastructure or workload provisioning.

As organizations adopt micro-segmentation, much more granular, application-specific security policies need to be deployed with the application. This further reinforces the importance of automation; manual processes fail as scale increases, especially for example at maintaining a production consistent environment throughout an Agile application development lifecycle.

The NSX REST API is leveraged by several Cloud Management Platforms (CMPs) to provide NSX services automatically (e.g., vRealize Automation, OpenStack, and VMware® Integrated OpenStack). vRealize Automation is of particular interest due to its tight integration with NSX as of NSX 6.0.

vRealize Automation provides native consumption of both pre-built and on-demand NSX network and security services. As of vRealize Automation 7.1 and NSX 6.2, organizations can provide end users the ability to automatically deploy a completely secure and compliant application topology utilizing NSX networking and security services. This is defined through vRealize Automation blueprints as shown in Figure 4.3.

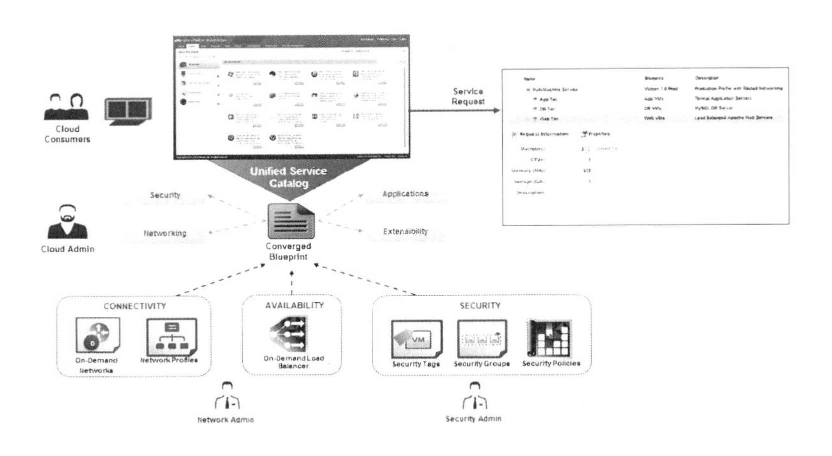

Figure 4.3 vRealize Automation and NSX

The Converged Blueprint Designer allows graphical design of end-to-end blueprints from virtual infrastructure up through a multi-tier application topology. For NSX, this includes creating:

- On-demand routed and NAT networks using network profiles
- Connections to pre-created external networks via existing NSX Logical Switches

- On-demand NSX logical switches and connecting them to pre-created NSX distributed logical routers
- An application-based security group with a default policy permitting traffic between tiers while blocking all inbound and outbound traffic for application isolation
- On-demand NSX security groups based on NSX security policies
- Membership in existing NSX security groups by specifying pre-defined NSX security tags
- An on-demand NSX load balancer in one-armed mode or in-line mode

This results in an integrated capability as shown in Figure 4.4. It also reinforces the need for blended teams as described earlier. Success with this level of integration and automation requires close cooperation between virtual compute, network, security, and storage resources.

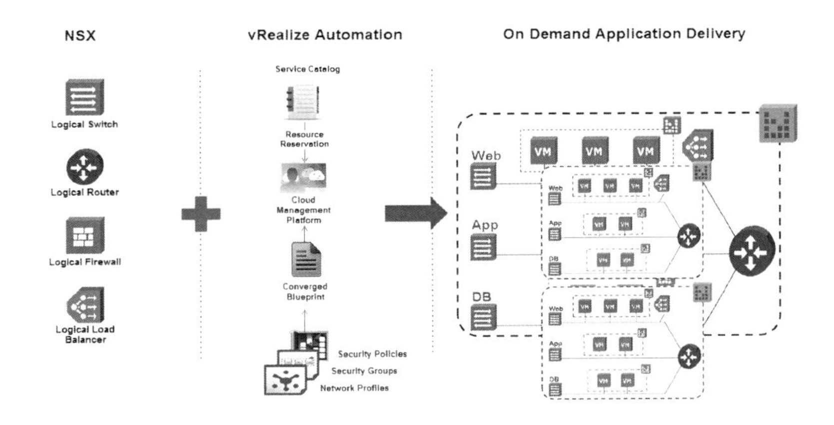

Figure 4.4 vRealize Automation and NSX integrated capability

vRealize Automation uses vRealize Orchestrator, allowing for out-of-the-box integrations using the Event Broker Service and XaaS features of vRealize Automation. The Event Broker Service allows subscribing vRealize Orchestrator workflows to specific events, modifying the behavior of the NSX native integration. XaaS allows publishing of vRealize Orchestrator workflows in vRealize Automation which can then be directly invoked from the vRealize Automation service catalog.

Blueprint-based integration between vRealize Automation and NSX allows deployment of full application stacks from the vRealize Automation service catalog. This provides users with the benefit of quickly deploying a fully configured, secure, and networked application stack within the framework of a standardized and repeatable process. An added benefit of using blueprints is the ability to treat infrastructure as code, allowing version control as part of an overall lifecycle management strategy.

KPIs

Table 4.5 Sample integrated NSX provisioning optimization KPIs

Objective	KPI	
Decrease average end-to-end workload provisioning time due to automating and continuously improving network and security-related steps.	• End-to-end monthly average workload provisioning time	This KPI should show a decrease in pre-provisioning and post-provisioning due to automating what had been manual network and security-related activities.
Increase workload provisioning success rate due to automating network and security-related steps.	• Monthly workload provisioning success rate • Monthly cost of provisioning failure	• Monthly workload provisioning success rate tracks total number of workloads requested, total number of workloads successfully provisioned, and total number of workload provisioning failures month over month which should reflect an increasing trend in the success rate. • Monthly cost of provisioning failure tracks the estimated or actual cost of workload provisioning failures which should decrease over time due to fewer failures.

Incident Management

Incident Management is also impacted in the context of operationalizing NSX. A number of tools and advanced capabilities can be applied to NSX; Chapter 6 - Tools for Monitoring & Troubleshooting will provide additional details on these possibilities. Some, like vRealize Log Insight, have content packs available to explicitly include NSX metrics and events. Others, like vRealize Operations Manager, can apply their advanced capabilities (e.g., dynamic thresholds, predictive analytics) to NSX components. Tools such as vRealize Network Insight have been purpose built for NSX to provide additional benefits (e.g., correlating physical and virtual network constructs). Used separately or together, these tools can significantly streamline troubleshooting and resolution across the software-defined network and security infrastructure.

NSX is software-defined with a REST API, easing integration with orchestration engines (e.g., vRealize Orchestrator) or intelligent operations tools (e.g., vRealize Operations Manager). This also allows for integration of NSX directly with 3rd party tools such as ServiceNow, an aspect particularly useful in the context of intelligent operations. A goal of intelligent operations is to identify and resolve issues before they impact a service or application. To help track the effectiveness of adopting an intelligent operations mindset, it is useful to track the ratio of issues resolved reactively (i.e., traditional incident management) against those resolved proactively. One strategy for this is to add the capability to mark entries in an ITSM tool's incident management module as "Resolved Proactively". vRealize Operations Manager can then invoke a vRealize Orchestrator workflow that creates an incident ticket with the pertinent information and is marked as "Resolved Proactively." This streamlines tracking the ratio of incidents resolved proactively to those resolved reactively for an NSX-based infrastructure.

Implementing the blended team model described earlier also streamlines incident management. A blended team acting collectively with shared knowledge is able to resolve issues much faster than the traditional model of siloed team hand-off and communication. This is especially valuable when there is pressure from a critical severity incident.

KPIs

Table 4.6 Sample incident management impact KPIs

Objective	KPI	
Decreased time to troubleshoot and remediate an incident due to blended team	• % of escalations resolved within the blended team • Average time to resolve escalations with the blended team	This KPI provides a measure of how effective the blended team is in resolving incidents.
Increase in the number of issues proactively resolved	• Number of NSX-related issues detected and resolved before they become service or application impacting incidents • % reduction in NSX-related incidents month over month	This KPI provides a measure of how effective proactive monitoring is in identifying and resolving issues before they become service or application impacting.

Compliance management

Isolation of application for security or regulatory compliance is a major use case for NSX micro-segmentation and specifically distributed firewalls. Integration of NSX distributed firewall rules with Active Directory allows for granular application access based on AD user groups. NSX distributed firewall rules can control application communication, allowing communication between application tiers while simultaneous restricting external inbound access based on port groups and other attributes. Distributed firewall rules are defined in NSX security policies and can be applied using NSX security groups. These security groups include all tiers associated with a multi-tier application, simplifying comprehensive application isolation. For an excellent treatment of NSX micro-segmentation - including designing and defining NSX security policies and security groups - see Wade Holmes' "VMware NSX Micro-segmentation: Day 1". This book can be downloaded from the URL provided in Table 7.1 in the "Where to go for more information" section.

Once security policies are created and applied to security groups, how are they managed on an ongoing basis? How best to monitor access and communication activity while supporting audit requests? There are several recommended solutions depending on the tools available in the environment.

At the most basic level, access log entries generated by distributed firewall events on a vSphere host, NSX Manager, or through the vSphere Web Client. The distributed firewall operations are run directly from the vSphere hosts.

Distributed firewall packet logs can be viewed locally on each vSphere host:

- Distributed firewall packet logs can be found at **/var/log/dfwpktlogs.log**
- Distributed firewall User World Agent (UWA) logs: **/var/log/vsfwd.log**

System events for distributed firewalls are accessed by downloading the tech support logs from NSX Manager administration GUI at Home -> Download Tech Support Log. This will generate a gzip file that can be downloaded for viewing/troubleshooting. Logs can also be sent to a remote syslog server (e.g., vRealize Log Insight) for easier analysis.

Audit logs associated with a specific vSphere host, including access control and firewall events, are viewed through the vSphere Web Client in the Networking & Security -> NSX Manager section. This interface can display raw audit log details as well as just the properties whose values have changed for a selected operation within an audit log.

The next level of distributed firewall auditing for compliance includes access monitoring. This can be done via the vSphere Web Client at Networking & Security -> Activity Monitoring. The following audit capabilities are available:

- User access (e.g., monitoring all VM access activity from Active Directory groups)
- Application access (e.g., monitoring access from VMs to a specific application)
- Inter-VM access (e.g., monitoring user- or service-to-application and VM-to-application access)

vRealize Log insight can perform more advanced distributed firewall auditing. It provides an easy yet powerful way to monitor and analyze distributed firewall rule events. A sample of out-of-the-box dashboards used to view summary-level distributed firewall events is shown in Figure 4.5.

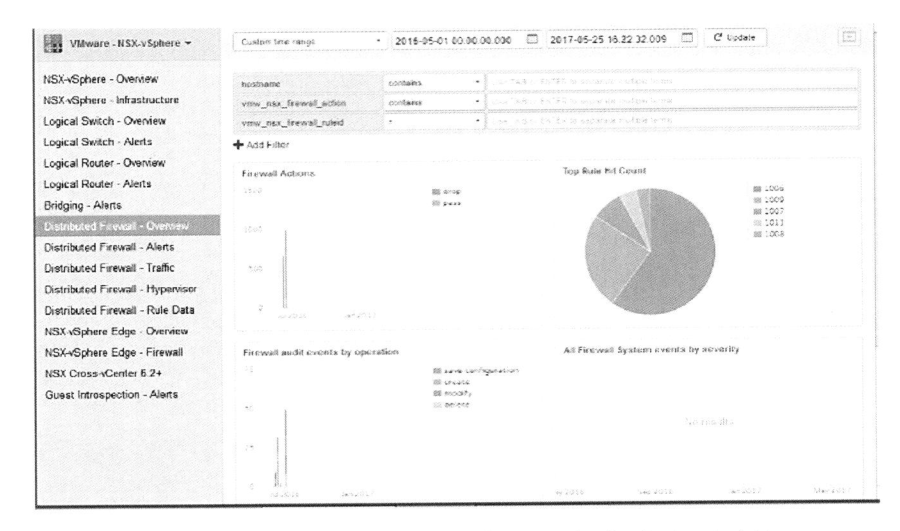

Figure 4.5 Distributed Firewall Event Summary in vRealize Log Insight

vRealize Log Insight also supports interactive analysis using logs in real time. In an example audit for access attempts shown in Figure 4.6, **172.16.60.22** (Web-03a) issued a ping to **172.16.60.12** (Web-04a) that was dropped due to FW rule # **1009**

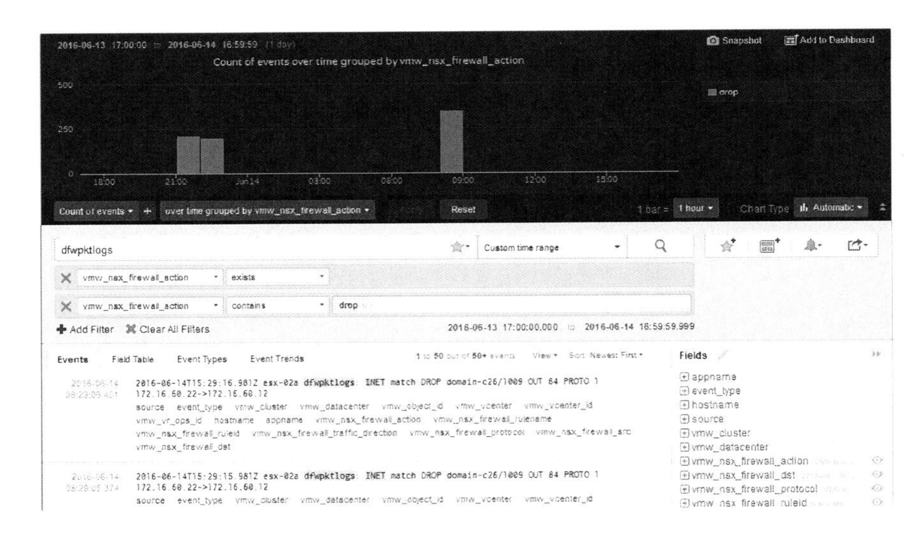

Figure 4.6 Auditing for access attemps using vRealize Log Insight

vRealize Network Insight can simplify NSX configuration as well as micro-segmentation compliance auditing. NSX configuration compliance can be accomplished using vRealize Network Insight's best practice checklist monitoring capability to validate NSX compliance against hardening guidelines. Figure 4.7 highlights the use of vRealize Network Insight for NSX micro-segmentation compliance to audit changes made to firewall rule membership over a specified period. To view this data, search for "firewall rule membership" and specify the date range for the changes.

Firewall Rule VM membership change

Firewall Redirect Rule: Prod_Midtier to Prod_Db Rule membership has changed. Added Lab-Midtier-5, deleted Lab-Midtier-10 — 154 days

NSX Firewall Rule: Prod MidTier to Prod Midtier Rule membership has changed. Added Lab-Midtier-5, deleted Lab-Midtier-10 — 154 days

NSX Firewall Rule: Lab to Prod Rule membership has changed. Added Lab-Midtier-5, deleted Lab-Midtier-10 — 154 days

NSX Firewall Rule: Prod Web to Prod MidTier - MidTierService membership has changed. Added Lab-Midtier-5, deleted Lab-Midtier-10 — 154 days

NSX Firewall Rule: Prod MidTier to Prod DB - DBService membership has changed. Added Lab-Midtier-5, deleted Lab-Midtier-10 — 154 days

NSX Firewall Rule: Prod to Lab Rule membership has changed. Added Lab-Midtier-5, deleted Lab-Midtier-10 — 154 days

Firewall Redirect Rule: Prod_Web to Prod_Midtier Rule membership has changed. Added Lab-Midtier-5, deleted Lab-Midtier-10 — 154 days

NSX Firewall Rule: Prod MidTier to Prod Midtier Rule membership has changed. Added Lab-Midtier-10, deleted Lab-Midtier-5 — 155 days

NSX Firewall Rule: Lab to Prod Rule membership has changed. Added Lab-Midtier-10, deleted Lab-Midtier-5 — 155 days

Figure 4.7 Firewall rule membership changes over time in vRealize Network Insight

The information identifies any firewall rule changes made directly or indirectly because of VM membership changes. This is critical to the audit change tracking process as it identifies why, when and how firewall rules changed. The changes can now be tracked, audited, and exported by following the live links.

This can be taken a step further by creating alerts to notify on changes, as shown in Figure 4.8, further augmenting the intelligent operations of the NSX environment.

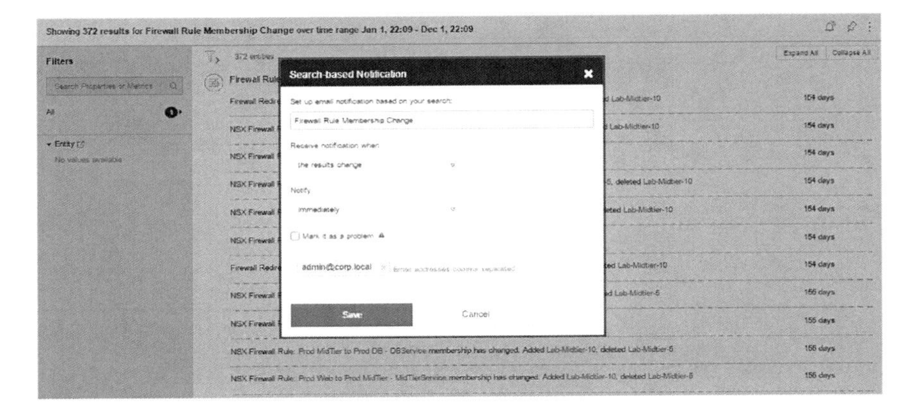

Figure 4.8 Creating change alerts for auditing in vRealize Network Insight

KPIs

Table 4.7 Sample compliance management impact KPI

Objective	KPI	
Zero instances of improper communication with or between applications in the NSX-based environment	• Average amount of time to detect and remediate improper communication with or between applications	This KPI provides a measure of the efficiency of identifying, validating, and remediating changes to application access or communication

Chapter 5

Consuming NSX

The end goal for operationalizing NSX is to support its use. This section provides high-level considerations and guidance for consuming NSX capabilities. This is approached from four perspectives:

- On-boarding organizations and users to consume NSX capabilities
- On-boarding applications to consume NSX capabilities
- On-boarding services to consume NSX capabilities
- Consumption of NSX capabilities by application and service developers

On-boarding consumers, applications, and services

Consumers

There are two models for envisioning the on-boarding of consumers: organizations requiring environment isolation (tenants in NSX documentation) and groups of users within an organization.

In the first model, a common example of an organization requiring environment isolation is the requirement for distinct production and dev/test environments on a common physical network. There should be no communication path between these two environments, but both may still require Internet access. This is represented in Figure 5.1.

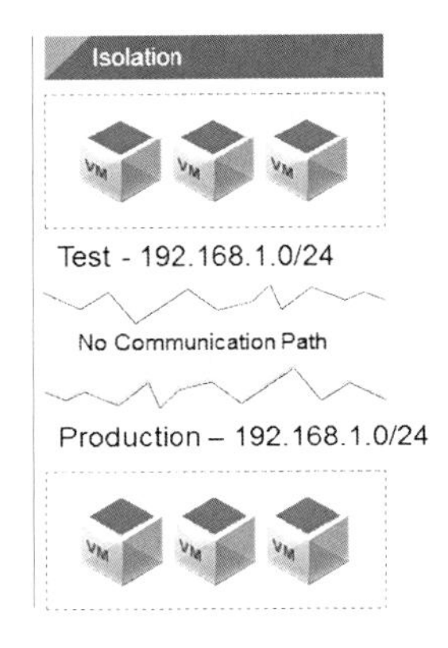

Figure 5.1 Separation between dev/test and production environments

The high-level steps for on-boarding an organization requiring isolation include:

- Create the primary VMware NSX Edge services gateway. This will act as the new organization's provider logical router (PLR) and logical switch that form the transit network connecting to the distributed logical router.
- Connect the primary NSX Edge services gateway uplinks to the external networks. Connect its internal interface to the transit network.
- Create the NSX for vSphere distributed logical router to provide routing capabilities for tenant internal networks. Connect its

uplink to the transit network.

- Create the secondary NSX Edge services gateway to handle the management traffic. Connect it to the management and transit networks.
- Create any known tenant networks and connect them to the NSX for vSphere distributed logical router.

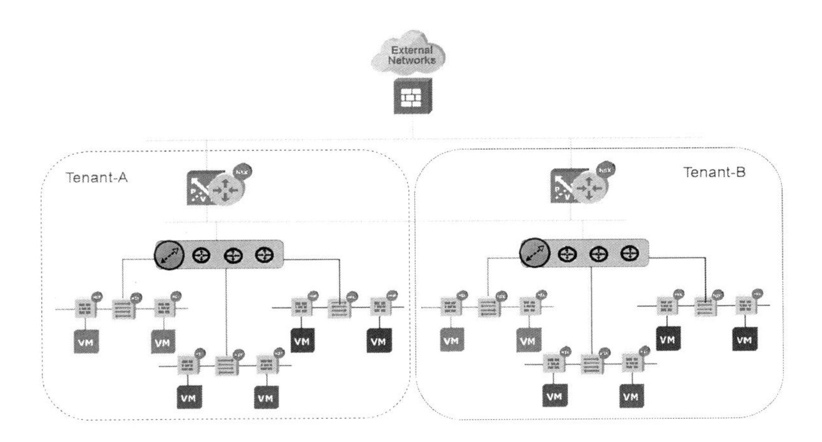

Figure 5.2 Isolated Tenant environments

A real-world customer use case is diagrammed in Figure 5.2, implementing the following components:

- External networks — These networks are external to the tenant. Connectivity to and from tenants is through the perimeter firewall. The main network under consideration is the Internet.
- Perimeter firewall — This is the physical firewall, which exists at the perimeter of the data center. Each tenant received either a full firewall instance or partition of a firewall instance to filter external traffic.
- Provider logical router (PLR) — The PLR is located behind the perimeter firewall and handles north-south traffic entering and leaving a tenant.
- NSX for vSphere distributed logical router (DLR) — The NSX for vSphere distributed logical router is optimized for forwarding in the virtualized space (i.e., east-west communication between VMs) on VXLAN- or VLAN-backed port groups.

- Internal non-tenant networks — This category is represented by a single management network located behind the perimeter firewall but not behind the PLR. The customer used this network to manage the tenant environments.
- Internal tenant networks — These networks represent the main tenant workloads that are connected to a distributed logical router that sits behind the PLR. These networks consist of VXLAN-based NSX for vSphere logical switches to which tenant virtual machine workloads will directly connect.

Adding groups of users within an organization is straightforward; add groups by registering one or more Windows domains with the NSX Manager and associated VMware vCenter® server. After registering the Windows domains, NSX Manager fetches group and user information from each domain. NSX Manager also retrieves Active Directory credentials associated with the Windows domains. Once NSX Manager has the Active Directory credentials, create security groups based on user or Active Directory group identity, create identity-based firewall rules, and run the activity monitoring reports discussed previously.

KPIs

Table 5.1 Sample on-boarding impact KPI

Objective	KPI	
Decrease the amount of time it takes to add a tenant	• Average time to add a new user or group to the NSX-based infrastructure • Average time to add a new tenant to the NSX-based infrastructure	These KPIs provide a measure of the effectiveness of automating the on-boarding of groups of users, and/or tenants

Applications

On-boarding applications is as much about setting up the NSX-based environment as it is about working with the application owners. Ideally, the development, test, integration test, user acceptance testing, and staging environments are also established. This is a critical capability that many organizations bypass. Assuming an application is already virtualized, the high-level steps for on-boarding an application include:

- Meeting with the application owner and architect to understand:
    - Overview of the application's architecture. Is it multi-tiered? Does it use containers or virtual machines? How tightly

coupled are the components? Does it require stateful or stateless communications?

    - Are there any security or regulatory compliance requirements? What security principles might apply? Are they company-wide or line of business specific? It is recommended to bring the security engineer from the blended team into the discussion.

    - What current security policies and firewall rules are applied to the application? Why are they in place? Can they be consolidated?

    - If the application is being placed in a segmented environment or if micro-segmentation is being applied to the application, is the target environment brownfield (i.e., existing) or greenfield (i.e., new)?

    - Are multiple environments required? Examples include development and test, quality assurance, user acceptance testing, pre-production staging, and production?

    - What are the communication requirements for external access to the application? What about communication with external systems or applications?

    - For multi-tier applications, what are the communication requirements between tiers?

- Design and configure any NSX networking services required for the application. Do any network-related SLAs or OLAs need to be taken into account?

- Design and configure new NSX security policies for the application

- Deploy the application and test environment to validate the NSX networking services and NSX security policies. Ideally the application already exists in a vRealize Automation blueprint or configuration management tool to simplify addition of NSX constructs and deploy from scratch. This helps maintain immutability (i.e., creation from original source) and idempotency (i.e., repeated deployment does not introduce change between environments).

- Once the application owner is satisfied that all required network and security functionality is in place, submit a change request for production deployment

- Deploy in production using a "blue/green" deployment model wherein both the new, NSX-based application and the original application run in parallel. Route some traffic to the new application for live validation before decommissioning the original application and its environment.

The line items "Design and configure any NSX networking services..." and "Design and configure any NSX Security Policies" can be critical, non-trivial activities depending on an application's complexity. To assist in this area, vRealize Log Insight and vRealize Network Insight can be used for micro-segmentation planning along with a new tool introduced in NSX for vSphere 6.3 - Application Rule Manager. Application Rule Manager provides a new way to quickly create security rulesets for new and existing applications. From a scalability perspective, it works extremely well to analyze and build larger rulesets quicker vRealize Log Insight, though it is still preferable to use vRealize Network Insight for large scale ruleset development. For a quick primer on using all three to develop micro-segmentation rulesets, review Geoff Wilmington's book *VMware NSX Micro-segmentation Day 2 Guide*, listed in Table 7.1 in the "Where to go for more information" section.

KPIs

Table 5.2 Sample application on-boarding impact KPI

Objective	KPI	
Decrease the amount of time it takes to on-board an application	• Average time to on-board a new application to an NSX-based infrastructure	This KPI provides a measure of a team's level of maturity in their understanding and use of NSX in on-boarding new applications

Services

Services in this context refer to user consumable services like Infrastructure as a Service, Platform as a Service, Digital Workspace as a Service, and Data Analytics as a Service. They may also refer to NSX-specific offerings (e.g., Load Balancer as a Service). On-boarding user consumable services starts with the definition of the service. A 360° service definition approach that includes all relevant stakeholders is recommended. These stakeholders could be from a line of business, IT, finance, marketing, or other organizations depending on the specific service. Involving all interested parties during the service definition dramatically increases the probability of success. This is more effective than traditional IT approaches of either collecting requirements and developing the service in isolation or developing a service without talking to the business.

Regardless of the approach taken, the process is comparable to that for on-boarding an application. If vRealize Automation capabilities are being used for service instantiation, map the service architecture and design resulting from the service definition to modular blueprints. vRealize Automation blueprints can be nested, so creating modular blueprints lends itself to potential reusability when creating other services. There is also the option of using user input to modify the VM and its environment as provisioned from the blueprint. User input such as application type or target production environment could dictate the security policies and security group assignment. NSX capabilities can also be exposed directly for end-user selection, such as the use of a load balancer.

With many options available when designing a service using a Cloud Management Platform (CMP) like vRealize Automation, developing a solid service definition is a critical starting point.

Developer consumption models

How do developers consume NSX capabilities when developing a new application? The recommended model is provisioning of virtual infrastructure from blueprints in which the application is developed.

Provisioning from blueprints represents the lowest risk approach. One of the most powerful aspects of using capabilities of blueprints (or Puppet manifests, Chef recipes, or Ansible playbooks) is the ability to provision the configuration defined in the blueprint repeatedly while guaranteeing the same result each time. This is ideal for application or service development by building on the base blueprint and ensuring the exact same configuration is provisioned when moving through the stages of the development lifecycle. In this way, the same configuration is provisioned in integration testing, user acceptance testing, staging, and production. Any changes should be made to the blueprint rather than directly in the environment with new environments deployed from these modified blueprints.

Developers interact with blueprints in one of two ways: selecting an entry in a service catalog, or deploying the blueprint via the API. Developers are increasingly opting for the latter approach; using REST API calls to deploy from a blueprint on-demand. A developer can use the vRealize Automation REST API to request a catalog item associated with a blueprint, then provide a JSON file containing values for any modifiable fields in the blueprint. This is a preferred approach for development teams who want the ease of using an API for provisioning a ready-made virtual infrastructure without requiring a granular level of infrastructure control.

Chapter 6

Tools for Monitoring & Troubleshooting

Monitoring and troubleshooting are the most visible and frequently asked about aspects of operationalizing NSX, and effective monitoring and troubleshooting begins with the tools. The NSX environment offers a wide variety of tools that enable proactive monitoring and provide effective troubleshooting. Figure 6.1 offers a framework for organizing the NSX tool ecosystem.

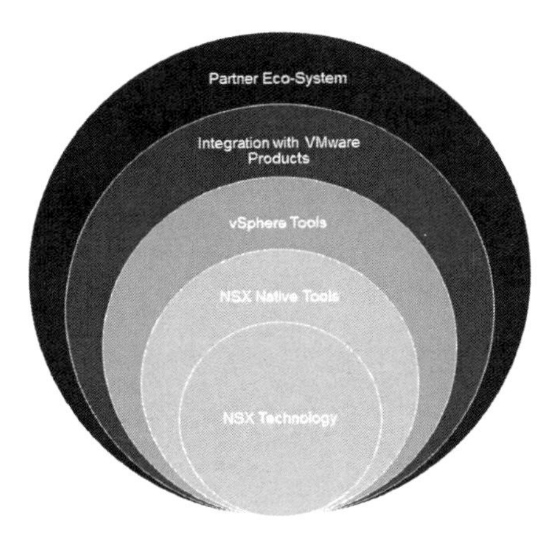

Figure 6.1 NSX tool ecosystem

These categories include:

- NSX Native Tools: troubleshooting capabilities delivered with NSX
- vSphere Tools: additional vSphere-based tools for NSX troubleshooting
- Tools provided by VMware Products: NSX monitoring and troubleshooting capabilities provided by the vRealize Operations, vRealize Log Insight, and vRealize Network Insight
- Partner Eco-System: additional NSX troubleshooting tools provided by 3rd party VMware partners

NSX Native Tools

NSX comes with a set of capabilities that provide extensive monitoring and troubleshooting capabilities. The primary built in tools for native monitoring and troubleshooting are accessed via the NSX Manager User Interface and the component command line interfaces (CLI). When coupled with component log files, these tools provide a standalone, comprehensive solution to monitor and troubleshoot an NSX-based infrastructure.

To improve troubleshooting, an NSX Central CLI was introduced in NSX 6.2. Previously, users had to log into each NSX component to retrieve information. The NSX Central CLI makes it easy to centrally compare information from the NSX Manager and other NSX components (e.g.,

NSX Controller, NSX Edge). Each NSX component can be configured via a single login, using the vCenter UI or using the REST API against NSX Manager. In a cross-vCenter NSX environment with multiple NSX Managers, a given NSX Manager can retrieve information from other NSX Managers about universal objects, while its own local objects remain private.

Table 6.1 provides information about key NSX native tools troubleshooting capabilities.

Table 6.1 Sample Native Tools troubleshooting capabilities

Operation	Tool
Logical network health	• NSX plugin to vCenter UI • NSX Manager Central CLI
NSX Controller	• NSX Manager Central CLI • NSX Controller CLI
Multi-level logging, event tracking, & auditing	• NSX Manager UI (also redirect NSX Manager, NSX Controller, NSX Edge syslog to remote syslog server) • NSX Manager Central CLI • NSX Edge CLI
Flow monitoring	• NSX Manager Central CLI • NSX Edge CLI
NSX routing	• NSX Manager Central CLI • NSX Edge CLI • NSX Controller CLI
Distributed Firewall	• NSX Manager Central CLI • NSX Edge CLI
Message bus	• NSX Manager Central CLI • NSX Edge CLI
Packet capture	• NSX Manager Central CLI • NSX Edge CLI • NSX Controller CLI
Load Balancer	• NSX Manager Central CLI • NSX Edge CLI
VPN	• NSX Manager Central CLI • NSX Edge CLI
All of the above	• NSX RESTful API

NOTE
For a comprehensive guide to NSX troubleshooting using the tools listed above, download the NSX Troubleshooting Guide referenced in Table 7.1 in the "Where to go for more information" section.

vSphere Tools

The next layer up is the NSX monitoring and troubleshooting toolset provided by vSphere. These tools are best accessed with the vSphere Web Client and vSphere CLI. The NSX Manager to vCenter initial connection process installs a web client plug-in for NSX on the vSphere Web Client server. NSX-specific capabilities are available from vSphere Web Client**> Networking & Security.**

Table 6.2 Sample vSphere-based troubleshooting capabilities

Operation	Tool
Overall health of NSX components	vSphere Web Client**> Networking & Security > Dashboard**
Detailed health of logical network	vSphere CLI (e.g., esxcli)
Status of communication between NSX components	vSphere Web Client**> Networking & Security> Installation > Host Preparation**
Performance issues	vSphere CLI (e.g., esxtop)
Packet flow	vSphere Web Client**> Networking & Security > Trace ow**
Packet visibility	vSphere Web Client**> Networking** (RSPAN/ ERSPAN)
Packet capture	vSphere CLI (e.g., pktcap)
NSX Distributed Logical Routing health	vSphere Web Client**> Networking & Security > NSX Edges**
Load Balancer debugging	vSphere Web Client**> Networking & Security > NSX Edges**

NOTE
For a comprehensive guide to NSX troubleshooting using the tools listed above, download the NSX Troubleshooting Guide referenced in Table 7.1 in the "Where to go for more information" section.

VMware vRealize Intelligent Operations Tools

The next level of the NSX monitoring and troubleshooting stack is VMware product tools. There are three VMware products for NSX monitoring and troubleshooting:

- vRealize Network Insight
- vRealize Operations
- vRealize Log Insight

vRealize Network Insight was purpose built for NSX monitoring and troubleshooting; it is the primary NSX monitoring and troubleshooting product for intelligent operations.

vRealize Network Insight

vRealize Network Insight is designed for intelligent operations of a software defined networking and security environment. It is an analytics tool focused on proactively enabling:

- Network health and performance monitoring
- End-to-end troubleshooting
- 360° visibility and analytics
- Micro-segmentation-based compliance management

The 360° visibility capabilities function across both the underlay (i.e., physical) and overlay (i.e., virtual) network fabric to troubleshoot and optimize network performance. This is based on selecting a source and destination object between which it provides visibility across both the virtual and physical layers. Figure 6.2 gives an example of an object's layer 3 and layer 4 path.

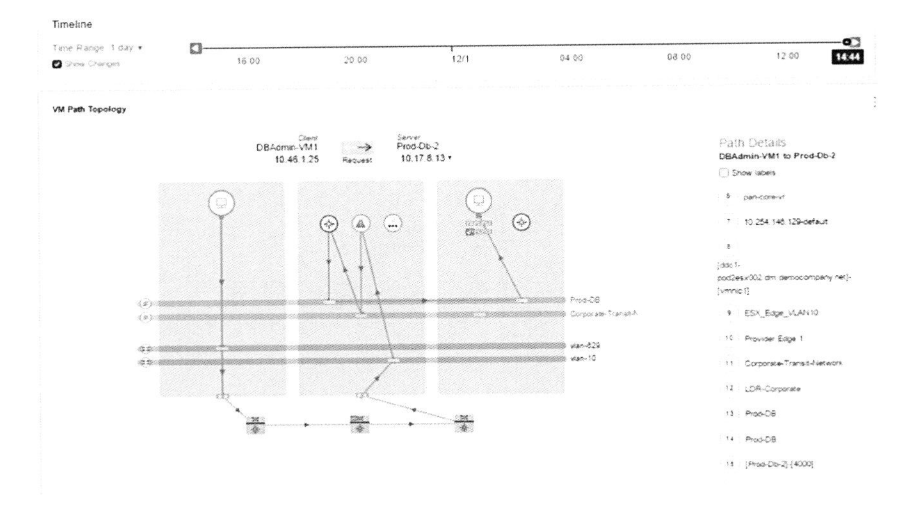

Figure 6.2 360° topology view

This provides a contextual view of the network covering each hop across VLANs and VXLANs. Key configuration information for each object is available through hover text., even for objects not managed by NSX. This configuration information is provided for everything in the path, both virtual (e.g., VM and host, VLAN and port group, VNIC) and physical objects (e.g., Cisco switches, Palo Alto Networks firewalls). Figure 6.3 shows a sample configuration for a Palo Alto Networks VRF.

Figure 6.3 Example Palo Alto Networks VRF configuration

In this example, traffic flows between objects as well as physical and virtual port path performance metrics. There is also a time machine feature to look at the state of the path between the selected objects at a given point in time. Objects allow drill down to view specific information about problems, changes, firewall rules, applicable flow paths, and localized topology.

vRealize Network Insight also excels at proactive micro-segmentation operations; to learn more about NSX micro-segmentation, download the *VMware NSX Micro-segmentation: Day 1* guide referenced in Table 7.1 in the "Where to go for more information" section. vRealize Network Insight provides a NetFlow-based assessment to model security groups and firewall rules. It also generates actionable recommendations for implementing micro-segmentation as well as monitoring micro-segmentation-based compliance over time. The NetFlow-based assessment capability uses real time analytics and flow data correlation to provide insights into the data traffic flow profile. For example, it can identify the percent of traffic that flowed east-west within the data center, what percent flowed to the Internet, what percent was routed through physical networks. It offers a breakdown of individual services and ports used over that 24-hour period, and can then make recommendations for firewall rule development based on real time communications. These recommended rules can be imported directly into NSX for refinement and implementation.

Closely related to micro-segmentation, vRealize Network Insight can monitor and troubleshoot security group-based compliance where NSX secures east-west traffic. vRealize Network Insight allows viewing of all deployed NSX security groups and associated data (e.g., associated VMs, data flows, firewall rules). Details are also available on add/delete/change events over time. Proactive monitoring alert definition allows for troubleshooting and remediation of issues before they impact security compliance.

vRealize Network Insight also provides purpose-built proactive health and availability monitoring, capabilities to understand problems and changes across all objects in the NSX environment, and custom event definition for proactive problem detection. This functionality is available for anything relevant to NSX, including VMs, ESX hosts, virtual networks, and firewalls. vRealize Network insight can also correlate problems on physical components with their virtual counterparts (e.g., mismatched MTU settings).

Another useful feature of vRealize Network Insight is the best practices checklist monitoring. The best practices checklist is a collection of NSX rules based on real-life deployments. These were developed from an operations' perspective related to the critical parameters and thresholds of the NSX management, control, and data planes. vRealize

Network Insight proactively monitors the NSX environment for violations of the best practice checklist rules and allows quick identification the offending components for remediation. A sample checklist failure screen is shown in Figure 6.4.

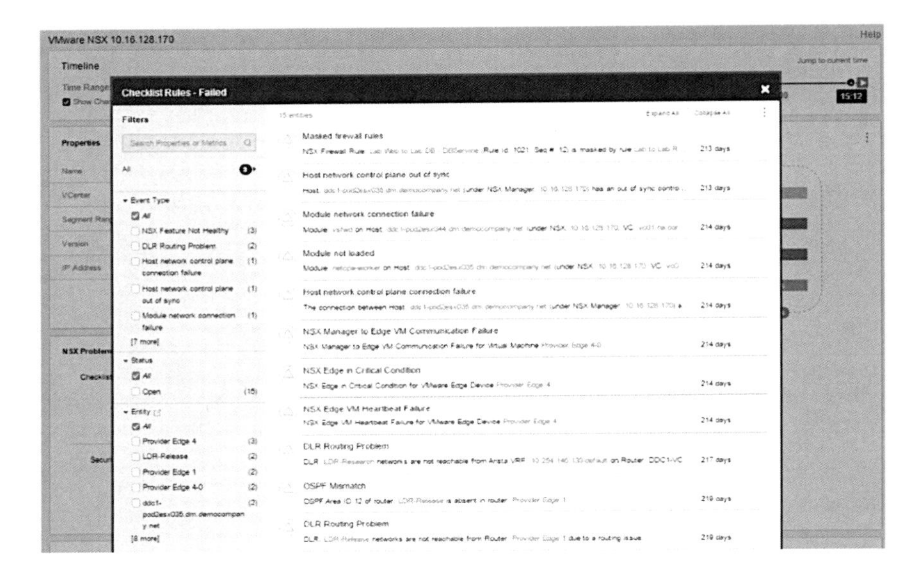

Figure 6.4 Example best practice checklist failure listing

This can be done through a rollup based on categorized problems, rollup based on affected objects as shown in Figure 6.5, or through a topology view of the NSX environment.

Figure 6.5 Example best practice checklist-based problem rollup

Drilling down through the rollup view provides access details about best practice checklist-based problems. The topology view includes a consolidated topology for cross-vCenter, multi-NSX Manager environments. The topology view allows for a quick drill down into NSX components - such as logical distributed routers or VXLANs - to see all the issues that have occurred based on violating the best practice checklist rules.

An example showing how checklist rules allow for quickly identifying problems comes from a customer who deployed vRealize Network Insight to monitor non-critical development and test environments as a pilot. The first problem involved conflicting firewall rules. As part of the daily routine reviewing checklist problems, they found a set of problems as shown in Figure 6.6:

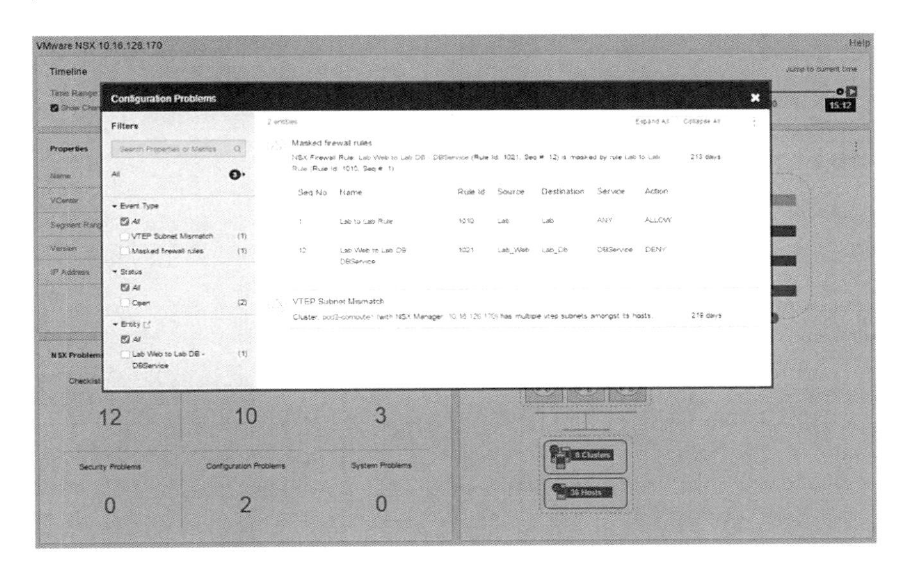

Figure 6.6 Customer best practice checklist-based problems detected

They could see the NSX controller is subjected to two firewall rules. The first rule is the **ALLOW**> **ANY** from **Lab to Lab**. This rule takes priority as Sequence Number **1.**

The second rule is Sequence number **12,** a lower priority that will not enforce the **Lab Web to Lab DB**> **DBService**> **DENY** rule. This was an easy fix to ensure Lab Web did not have direct access to the DBService, as they only wanted access through their middleware tier. This example illustrates proactive operations; identifying conflicting firewall policies and resolving them before they become service impacting. In this case, it is preventing a developer from potentially writing code assuming direct access to a database service that would not be available in production.

vRealize Operations

vRealize Operations provides intelligent operations from application to infrastructure across physical, virtual, and cloud environments. It compliments vRealize Network Insight, allowing use of all its capabilities to predictively manage NSX components as virtual machines or virtual appliances; for example, applying dynamic threshold-based intelligent analytics to an NSX Edge cluster as a collection of virtual machines based on a custom tag. vRealize Operations offers smart alerts, guided root cause analysis, and automated remediation to identify and resolve issues in the NSX Edge cluster from a VM perspective.

vRealize Log Insight

vRealize Log Insight provides highly scalable log management with intuitive actionable dashboards, sophisticated analytics, and broad third-party extensibility. The Log Insight Content Pack for NSX provides operational reporting, trending, and alerting visibility for all sources of log data within NSX. Each major NSX function - logical switching, routing, distributed firewalls, VXLAN gateways, and edge services - is represented via custom dashboards, filters, and alerts. This content pack provides last mile troubleshooting to compliment the monitoring and troubleshooting capabilities of vRealize Network Insight and vRealize Operations.

It is structured in a hierarchical manner, starting at an overview level showing problems based on a rollup of underlying NSX components. Each level provides specific, actionable information associated with the problem. This guides the troubleshooting effort to the appropriate next level of NSX component dashboard that shows specific alerts. These alerts display specific problems and recommended remediation actions.

Figure 6.7 shows the out-of-the-box top dashboard for NSX within vRealize Log Insight.

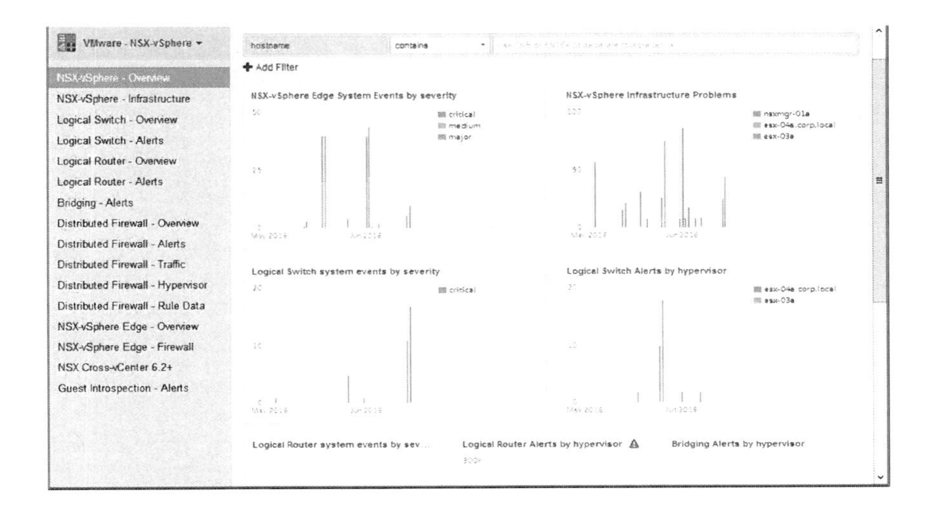

Figure 6.7 NSX Overview dashboard in vRealize Log Insight

Selecting **Distributed Firewall – Overview** results in Figure 6.8.

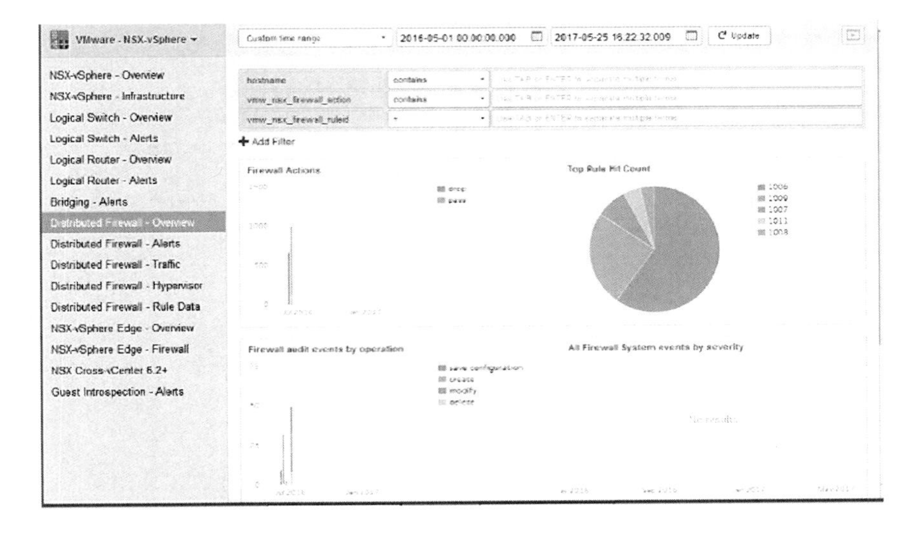

Figure 6.8 Distributed Firewall Overview dashboard in vRealize Log Insight

To further explore the **Firewall Actions** resulting in drops, hover over the line depicting drops in the **Firewall Action** pane and access the details shown in Figure 6.9.

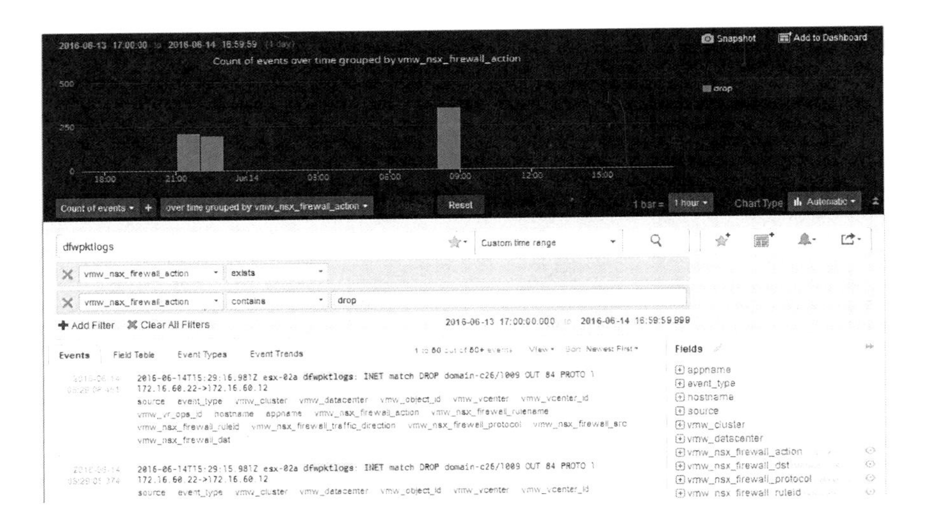

Figure 6.9 Firewall Actions log details in vRealize Log Insight

The log entry shows **172.16.60.22** (Web-03a) issued a ping to **172.16.60.12** (Web-04a). It was dropped due to FW rule # **1009**, which is the expected behavior.

Figure 6.9 also demonstrates vRealize Log Insight's filtering power. The **Filter** allows a user to dynamically extract any field from the data using regular expression. The extracted fields can be used for selection, projection, and aggregation. The **Fields** pane on the right side of the screen allows for customization in search and display of data classifications and keywords.

Partner Ecosystem Tools

NSX has a growing ecosystem of technology partners that have integrated monitoring, troubleshooting, logging, and auditing functionality. For the latest list of tools, please reference table 7.1 on page 72.

Chapter 7

Conclusion

As part of VMware's Software Defined Data Center, NSX delivers a previously unheard-of level of flexibility and agility with its software-defined networking and security capabilities. To fully leverage these capabilities in a sustained manner, VMware highly recommends optimizing operations for a software-defined infrastructure. To be truly effective, optimize the deployment of the most valuable resources – people – in blended teams. Ensure they have the right skills to not only make the NSX operations successful, but to make them successful as individuals and as a team. Optimize critical operational processes to take advantage of the opportunity software defined networking and security provides. Leverage the new breed of tools providing intelligent operations available to enhance monitoring and troubleshooting capabilities for NSX and the software defined datacenter. Making these changes enables an operating model that allows a shift to an intelligent operations mindset. Adopting NSX technology and adapting operating models in these ways will best leverage the investments in and realize the benefits of software-defined networking and security.

Where to go for more information

Description	Reference
Organizing for the Cloud	https://www.vmware.com/content/dam/digitalmarketing/vmware/en/pdf/solutionoverview/vmware-oganizing-for-the-cloud-feb2017.pdf
NSX Troubleshooting Guide	https://docs.vmware.com/en/VMware-NSX-for-vSphere/6.3/com.vmware.nsx.troubleshooting.doc/GUID-22AA06B4-2AA7-4A23-8AF8-D2D81CB72FBA.html
Automation Leveraging NSX REST API	https://communities.vmware.com/docs/DOC-31921
NSX documentation	https://www.vmware.com/support/pubs/nsx_pubs.html
VMware NSX Micro-segmentation: Day 1	www.vmware.com/go/run-nsx
VMware NSX Micro-Segmentation Day 2	www.vmware.com/go/run-nsx
Relevance in the Age of Cloud	https://www.vmware.com/content/dam/digitalmarketing/vmware/en/pdf/whitepaper/achieve-relevance-in-age-of-cloud.pdf
High Performance Organizations	"Lean Enterprises," by Jez Humble, Joanne Molesky, and Barry O'Reilly
VMware NSX Technology Partners	https://www.vmware.com/products/nsx/technology-partners.html
For NSX CLI, API, Logging & System Events	https://docs.vmware.com/en/VMware-NSX-for-vSphere/index.html

Table 7.1 Reference

Index

A

Active Directory 42, 43, 52
Agile 1
API 22, 37, 38, 41, 55, 59
Application-centric 30, 32
Application Rule Manager 54
Audit 23, 37, 42, 43, 44, 45

B

Blended team 13, 14, 15, 16, 17, 18, 19, 24, 25, 26, 36, 41, 42, 53
Blueprint 15, 34, 35, 37, 53, 55
Business stakeholders 9, 11, 25, 35

C

Cloud
- *Cloud Architect* 17
- *Cloud Automation & Integration Developer* 19, 20, 21, 22, 23
- *Cloud Management Platform* 55
- *Cloud Network Administrator* 23
- *Cloud Network Architect* 18, 20
- *Cloud Network Engineer* 20, 23
- *Cloud Security Administrator* 24
- *Cloud Security Architect* 19, 21
- *Cloud Security Engineer* 21, 24

Compliance management 42
Configuration management 36
Converged Blueprint Designer 38
Cross-domain 13, 14, 15
Cross-functional 13, 14, 15, 26
Culture 3, 25

D

Digital transformation 2
Distributed firewall 33, 42, 43

E

Event Broker Service 40

I

Incident Management 41
Intelligent analytics 31, 66
Intelligent Operations 4, 29

M

Mindset 2, 3, 11, 25, 30, 32, 41
Monitoring and troubleshooting 4, 57, 58, 60, 61, 66

N

NSX
- *NSX Central CLI* 58
- *NSX Edge* 31, 32, 33, 50, 51, 59, 66
- *NSX Native Tools* 58
- *NSX Troubleshooting Guide* 60

O

Operating model optimizations 11
Operationalizing NSX 2
Operational processes 2, 3, 4

P

People 2, 3, 8, 11, 12, 25
People Considerations 4, 8, 11
Performance and availability 30, 31, 32
Plan, build, and run 12, 13, 15
Proactive monitoring 30, 63
provisioning 15, 18, 19, 22, 35, 37, 40, 55

R

REST API 37, 38, 41, 55, 59
Roles and skillsets 3, 24

S

Security
  Security group 39, 55, 63
  security policy 16, 21
Software Defined Datacenter 1, 13
Software defined networking and security 1, 2, 11, 14, 15, 17, 25, 26, 29, 30, 32, 61

T

Team structure 3, 13, 26
Tiger team 8, 15, 16, 17
Tooling 2, 3

V

VMware
  NSX
    NSX Central CLI 58
    NSX Edge 31, 32, 33, 50, 51, 59, 66
    NSX Native Tools 58
    NSX Troubleshooting Guide 60
  vRealize
    vRealize Automation 34, 36, 38, 39, 40, 53, 55
    vRealize intelligent operations tools 29
    vRealize Log Insight 21, 23, 24, 37, 41, 43, 44, 54, 58, 61, 66, 67, 68
    vRealize Network Insight 21, 23, 24, 30, 31, 41, 45, 46, 54, 58, 61, 63, 65, 66
    vRealize Operations Manager 31, 41
    vRealize Orchestrator 22, 36, 40, 41
  vSphere
    vSphere Tools 58, 60
    vSphere Web Client 42, 43, 60

X

XaaS 40